T0305618

Geostatistics Notes for Practitioners

This book provides a practical perspective of all the processes involved in estimating mineral resources and reserves, including mine-to-mill reconciliation. It provides an integrated step-by-step explanation of processes for performing each step, including insight from academic and industry practitioners. Each chapter details a specific aspect of the estimation processes in a practical manner. It contains examples and case studies to illustrate the practical application of geostatistics in mineral resource estimation, mineral reserve conversion, and reconciliation.

Features

- Provides a step-by-step guide with over 10,000 lines of Python code for hands-on demonstration, from start to finish, for both linear and non-linear geostatistical methods.
- Explains practical geostatistics processes and functionality.
- Simplifies explanation of mathematical/statistical concepts and application.
- Discusses generalised examples to aid the process steps.
- Reviews processes involved in the mineral resources' estimation and ore reserve conversion.

This book is intended for third-year and postgraduate students in Mineral Resources Management, Geology, Spatial Statistics, and Mining Engineering, as well as practising professionals.

Geostatistics Notes for Practitioners

Glen Nwaila, Leon Tolmay,
and Mark Burnett

CRC Press
Taylor & Francis Group
Boca Raton London New York

CRC Press is an imprint of the
Taylor & Francis Group, an **informa** business

Designed cover image: © Shutterstock, Oyls

First edition published 2025
by CRC Press
2385 NW Executive Center Drive, Suite 320, Boca Raton FL 33431

and by CRC Press
4 Park Square, Milton Park, Abingdon, Oxon, OX14 4RN

CRC Press is an imprint of Taylor & Francis Group, LLC

ISBN: 978-1-032-59926-7 (hbk)
ISBN: 978-1-032-65045-6 (pbk)
ISBN: 978-1-032-65038-8 (ebk)

DOI: 10.1201/9781032650388

Typeset in Times
by codeMantra

Access the Support Material: www.routledge.com/9781032599267

"Knowledge is not a stagnant reservoir to be filled, but a dynamic spark that ignites a lifelong journey of discovery."

We are humbled to have written the practitioner's notes in the same halls of academia at the University of the Witwatersrand (South Africa), where the Father of Mining Geostatistics, Prof. Danie G. Krige, once walked and inspired many. Through his teachings, we have learned that the true path to mastering mining geostatistics lies in constantly pursuing knowledge and applying that knowledge in practical settings. With this in mind, we offer these practitioner's notes as a tribute to the unnamed miners who have dedicated their lives and abilities to unlocking the Earth's secrets.

Contents

Foreword

The wealth of mineral resources that lies beneath the Earth's surface drives technological advancements and underpins our economic prosperity. This comprehensive book presents the methods and processes of mineral resource estimation at each step along the mining value chain and provides a necessary framework for understanding mineral development and exploitation. The complexities associated with mineral resource estimation are addressed by explaining the importance of reliable methodologies and the consideration of uncertainty in decision-making.

The challenge of determining the true in situ value of mineral deposits requires insight and understanding of a wide range of disciplines which are laid out in an understandable manner in this volume. The tools used open a path for its readers along a stepwise process that produces a block valuation of a mineral deposit as well as providing the means for quantifying the uncertainties associated with the evaluation.

The numerous sources of risk associated with capital-intensive investment in mineral development and mining demand meticulous consideration of the nature and size of uncertainties and risks at every stage. From exploration to the end of the life of mine, the critical need for risk quantification at every stage of the mining value chain as captured in the distinction between measured mineral resources or proved mineral reserves and their inferred level estimates is highlighted in this volume.

The volume examines the unique characteristics of mineral deposits such as ore paragenesis, geographical location, socio-economic and political settings, and environmental considerations and describes the way in which they shape the viability of mining projects. This understanding lays the groundwork for risk mitigation and informed decision-making throughout the resource estimation process.

The history of geostatistics from its early beginnings to its widespread use in present-day mineral resource estimation is examined. By combining statistical analysis with spatial data, geostatistics has provided insight and understanding of critical importance to the mining industry as well as natural phenomena across a wide range of other disciplines. The pioneering work of Danie G. Krige and Georges Matheron led to the later development of methods such as multiple-point statistics and geostatistical simulation, emphasising the enduring relevance of this remarkable algorithm.

Sampling, a fundamental component of mineral resource estimation, takes centre stage in this volume. The principles underlying the sampling process are examined, covering a spectrum of methods from conventional grab sampling to advanced systematic sampling. Emphasis is placed on the Theory of Sampling, the precursor for representative sampling, ensuring that collected samples faithfully represent the mineral deposit's heterogeneity and intrinsic characteristics.

Geological factors form the critical building blocks for all stages of mineral resource estimation. The authors explore the key geological factors influencing mineral resource estimation, highlighting the importance of a comprehensive geological understanding of the deposit. Addressing geological uncertainty through probabilistic modelling and geostatistical simulations, this chapter underscores the significance of interdisciplinary collaboration between geologists, resource estimators, and mining engineers.

The book lays out the essential preparatory steps for the estimation process. Thorough data review, quality assurance, and integration of geological understanding are emphasised. The description extends to exploratory data analysis, variography, kriging, conditional bias, and cross-validation, providing a foundation for accurate regionalised variable estimation.

The discussion on the classification of mineral resources followed by their conversion to economically viable mineral reserves underscores the complexity and importance of post-mineral resources estimation processes. The section explains how accurate classification and understanding of dispersion variance contribute to sustainable and profitable mining ventures.

The chapter on nonlinear geostatistical methods and their application in the conversion of mineral reserves provides a comprehensive understanding of these advanced techniques. The discussion equips readers with the knowledge and tools necessary to improve mineral reserve classification and conversion processes, ultimately leading to informed and reliable decision-making in the mining industry.

Mineral resources management and mine planning are emphasised in the chapter on grade control. It underscores the critical role of correctly separating ore and waste in ensuring sustainable mineral extraction practices, offering valuable insights for enhancing ore grade evaluation, risk management, and the success of mining projects.

Reconciliation, the art of harmonising divergent concepts or facts in the mining industry is the focus of the subsequent chapter. The chapter discusses the challenges and solutions in implementing mine-to-mill reconciliation strategies, fostering a culture of collaboration and data-driven decision-making within mining companies.

This book concludes with an exploration of geostatistical methods for accurately estimating recoverable mineral resources. The significance of change-of-support techniques in enhancing the accuracy of mineral resource estimates is highlighted, offering readers a comprehensive understanding of the latest methodologies and tools for mineral resource estimation.

In conclusion, this book serves as a comprehensive guide to mineral resource estimation and the mining value chain, offering valuable insights and practical strategies for professionals, researchers, and students in the mining industry. By explaining the intricacies of geostatistics, sampling, geological considerations, and advanced estimation techniques, this book equips readers with the knowledge and tools necessary to navigate the complexities of mineral resource estimation and contribute to the sustainable and responsible management of mineral resources worldwide.

Richard Minnitt, Em Professor, School of Mining Engineering,
University of the Witwatersrand

FOREWORD...GLEN NWAILA'S BOOK (WITH LEON TOLMAY AND MARK BURNETT)

Geostatistics notes from practitioners: A practical guide for estimating mineral resources.

The mining industry has witnessed its fair share of scandals and misrepresentation over the centuries – unsurprisingly – because by their very nature, ores are obscure entities that are difficult to find, frequently hidden by overburden and not easily differentiated from the rock sequences that host them. The measurement of the grade of an ore deposit is fraught with difficulties, related to the fact that some minerals occur in vanishingly small amounts and the ores themselves are spatially distributed in a haphazard way that often defies logic and system. The importance of geostatistics to the minerals industry, and the need for rigorous procedures by which the quantification and reporting of mineral deposits can be made, has long been recognised – especially after the spectacular revelations that arose, for example, from the notorious Erfdeel gold-salting case, so elegantly described by Dr Joe Liebenberg in his Presidential Address to the Geological Society of South Africa in 1961.

For this reason, there has been an urgent need to develop methods by which the quantity of a mineral in a rock can be estimated and the spatial variability of this quantum described – these need to be accessible to all interested parties ranging from the geologist logging a section of borehole core to an investor wishing to sink his last dollar into a mine development. And so the science of GEOSTATISTICS was born. This is a mathematical discipline in which conventional statistical techniques are applied to an estimation of the size of an orebody and the spatial distribution of metals contained within it.

The roots of geostatistics are well known – the pioneering scientists who developed this discipline are recognised worldwide. Geostatistics developed rapidly from around the 1960s due to the work of a seminal few, and it was concentrated mainly in two countries – South Africa and France. Names such as Danie Krige, Herbert Sichel, and Georges Matheron are now cast in the pantheon of legendary mathematical-statistical pioneers. It is therefore appropriate that this new book *Geostatistics Notes for Practitioners* by Glen Nwaila, Leon Tolmay, and Mark Burnett should be rooted in South Africa, and in particular at the University of the Witwatersrand.

It is also appropriate that geostatistical methods are increasingly popularised, distributed, and utilised worldwide. As we recover from a global pandemic and economic downturn, we enter the 'Green Revolution' in which people are demanding a lifestyle that is no longer entirely dependent on fossil fuels. Climate change is an existential problem that will be solved in part by the implementation of renewable energy strategies and the exploitation of an entirely new suite of mineral deposits. Future best practice revolves around the responsible management of our precious resource base, and this book goes a long way to ensuring that this is done in a scientifically rigorous manner – one that lays the groundwork for responsible and ethical management of global mineral resources.

Prof. Laurence Robb, University of Oxford

Preface

Many books and publications have been produced in the discipline of geostatistics detailing the various mathematical and statistical methodologies that underpin the science, as well as one or two that attempt to explain the rigour required when applying those principles in reality. A primary objective of this book is to provide a comprehensive explanation of the practical aspects that must be understood by persons undertaking the geostatistical estimation of a mineralised deposit. The authors attempt to provide practical methods of achieving the desired results by detailing the required checks and balances before proceeding to the next stage of a geostatistical estimate. Additionally, examples of how the authors manage to accomplish their goals with a limited amount of hassle and specialised software will be provided. This is not to say that the authors have neglected the mathematical basis of geostatistics; instead, these will be limited as much as possible. What is attempted is to provide practical methods to achieve the required technical framework to achieve a 'correct' estimate. The authors will explain why specific processes must be accomplished, what their implications are, what the underpinning mathematical models are, and how they are applied.

If you, dear constant reader, are anticipating a text that will default to a mechanistic stepwise approach of calculation, semi-variogram modelling, and the application of Kriging on a 30 m × 30 m grid (for some reason, the industry default size, which likely originates from Danie Krige's very early framework of geostatistics as well as the practice of having a panel or stope length of 30 m with a face advance of 10 m per month) using a minimum of two samples and a maximum of six samples, there you have a robust estimate of your deposit. If this is your anticipation, then alas, constant reader, you are doomed to disappointment, as, in the authors' experience, the estimation of mineral resources requires a multiplicity of steps to be successfully applied in order to produce the desired results, in our case a valid estimate of the contained metal, grade distribution, and tonnages contained within our mineralised deposit. The authors start with the principles needed to be entrenched in your geostatistical practice before a single sample is taken or borehole drilled. You will then be guided through all the checks and controls required that will determine if your data are robust and reliable and will not cause a bias in your estimate. From here, we will progress into how we convert data into information gathered from various sources including mining, rock engineering, planning, geology, finance, and metallurgy. Once all of these are obtained, we can now start the pre-estimation analysis and checks. Once all of these have been completed, you can start determining the methodology to estimate the deposit. This, in turn, requires that additional data be collected and analysed before one can attempt and finally attempt evaluation (i.e., grade and tonnage estimation). However, this is not the end of our journey.

Once we have completed our estimation, we are required to validate it; to do so, a series of post-estimation processes have to be completed to ensure that the final estimate results are in a form that can be utilised for resource evaluation and mine planning. The implicit assumption that the authors are making is that by diligently following the processes and checks outlined in this text, the final result will be a mineral resource and mineral reserve that can be reported following the requirements of one of the reporting codes or standards that are aligned with the CRIRSCO International Reporting Template for the Public Reporting of Exploration Targets, Exploration Results, Mineral Resources, and Mineral Reserves, November 2019 (the 'CRIRSCO International Reporting Template 2019') developed by the Committee for Mineral Reserves International Reporting Standards (CRIRSCO). These codes and standards include JORC (Australasia), CBRR (Brazil), CIM (Canada), Comision Minera (Chile), CCRR (Colombia), PERC (Europe), NACRI (India), KCMI (Indonesia), KAZRC (Kazakhstan), MPIGM (Mongolia), OERN (Russia), UMREK (Turkey), SAMCODES (South Africa), and PMRCC (Philippines). In conclusion, the authors hope that this text will benefit the reader, transferring the author's experience to the next generation and the current generation of practitioners.

About the Authors

Glen Nwaila is an Associate Professor at the School of Geosciences and former Director of the Wits Mining Institute, University of the Witwatersrand (Wits). Glen joined Wits in the fall of 2017 and was subsequently appointed an academic staff member. Prior to joining Wits, he worked in the mining and consulting industries, where he led teams of mining professionals and led audits in the mineral resources and extractive metallurgy plants. At Sibanye-Stillwater and Harmony Gold, Glen led the geology functions in several mining and metallurgical operations. He also served as Manager at Deloitte Technical Mining Advisory. Glen is an Erasmus Mundus Scholar and Visiting Professor at Uppsala University (Sweden) since 2018. Glen completed his PhD with Magna Cum Laude at the Julius-Maximilians-Universität Würzburg in Germany, MSc degree in Chemical Engineering from the University of Cape Town, and an Honours degree in Geology from the University of Johannesburg. He currently serves as one of the 35 IGC Legacy Fund's Board of Directors. He is a Professional Natural Scientist with the South African Council for Natural Scientific Professions (SACNASP) and is a Fellow of the Geological Society of South Africa (FGSSA). Glen's research focuses on improving ore characterisation and modelling to enhance selective mineral resource extraction, minimise environmental impacts, increase profitability, and maximise natural resource utilisation. Specifically, Glen and his students work on geometallurgy, machine learning, and spatial data analytics research projects related to multiscale and multivariate data integration, big remote sensing, and geochemical data analysis to enable optimal decision-making in the presence of uncertainty. Glen's research interests are (1) the genesis and evaluation of ore deposits, (2) machine learning applied to geoscience and extractive metallurgy, (3) metal accounting, and (4) process optimisation in hydrometallurgical plants. The emphasis of his most recent work has been on the impact of digital transformation in the mining industry and the move toward dry labs and data banks for process simulation and gamification.

Leon Tolmay is the Founder and Managing Director of Tolmay Enterprises and an associate of SmartMin. He has more than 42 years of practical geostatistics in the mining and metals industry. Before founding Tolmay Enterprises, Leon served as Evaluation manager in charge of Geostatistics and Evaluation for the Sibanye-Stillwater Gold Division, responsible for all resource and reserve modelling and classification. Prior to joining Sibanye-Stillwater, Leon was a lead geostatistics consultant for Goldfields. He also worked as a mining surveyor, Chief mine surveyor, mine planner, mineral resource manager, ventilation officer, and TSD officer, as well as during a very short period during a wild cat strike a miner. Throughout his career, Leon has created several geostatistics training guides and taught geostatistics as a guest lecturer in institutes of higher learning. Leon spent the early years of his career under the mentorship of Prof Danie Krige (School of Mining Engineering, University of the Witwatersrand) together with colleagues Dr Winfred Asasibey-Bonsu (Group Geostatistician and Evaluator, Goldfields, Australia).

Mark Burnett has more than 30 years of experience in the mining industry, including mineral exploration, shaft sinking, managing the geological and mineral resource estimation functions on operational mines, mergers, acquisitions, and asset disposal. Mark is currently part of the Pan European Reserves and Resources Reporting Committee's (PERC) Executive and is one of PERC's representatives on the Committee for Mineral Reserves International Reporting Standards (CRIRSCO). Mark is currently an active member of three of the United Nations Resource Management System's (UNRMS) Expert working groups. Mark has an MSc degree in Mineral Resource Management from the University of the Free State (South Africa) and an Honours degree in Geology from the University of the Witwatersrand (South Africa). Mark is a registered Chartered Geologist (CGeol) with the Geological Society of London and a registered European Geologist (EurGeol) with the European Federation of Geologists.

Acknowledgements

We extend our heartfelt gratitude to the individuals whose invaluable contributions and unwavering support made the creation of this book possible. Their dedication and expertise have undoubtedly elevated the quality of this work, and we are honoured to express our sincere appreciation to each one of them.

First and foremost, we are deeply grateful to Dr Derek Rose for his exceptional skills in graphic design. His artistic flair and keen eye for detail have brought life to the figures and illustrations within this book, enhancing its visual appeal and comprehension. His commitment to excellence and willingness to go above and beyond have been instrumental in shaping the final presentation of the content.

We would like to express our profound thanks to Dr George Henry, whose insightful editing has played an important role in refining the clarity and coherence of this book. His meticulous attention to detail and constructive feedback have significantly improved the overall flow and readability of the text. His support and encouragement throughout the editing process have been invaluable.

A special word of gratitude goes to Phumzile Nwaila for her dedicated efforts as the proofreader of this book. Her meticulous eye for grammatical errors, spelling mistakes, and typographical discrepancies has ensured that the text is polished to perfection. Her commitment to maintaining the integrity of the content is commendable.

We are grateful to Dr Steven Zhang and Julie Bourdeau for assisting in various aspects of proofreading this book.

We would like to acknowledge the significant contributions of Dr Nelson Chipangamate, who not only played a key role in formatting this book but also contributed his insightful suggestions during the editing phase. We also extend our gratitude to Prof. Megan Becker, Prof. David Reid and Prof. Hartwig Frimmel for their guidance during the early stages of conceptualization and drafting of this book.

Last but certainly not least, we extend our sincere appreciation to Prof. Laurence Robb and Prof. Richard Minnitt for their invaluable support in proofreading the manuscript and for graciously writing the foreword. Their expertise in the field and his willingness to lend his time and expertise to this endeavour have been truly humbling.

In addition to the individuals mentioned above, we would like to express our gratitude to all those who have supported us in various ways throughout this journey. To our families, friends, and colleagues, your encouragement and belief in us have been a constant source of motivation.

Finally, we would like to acknowledge the readers of this book. It is our hope that this work proves insightful, engaging, and of value to you. Any success achieved through this book is undoubtedly a result of the collective effort and dedication of everyone mentioned here and many others who have played a role behind the scenes.

Abbreviations

BLUE	Best Linear Unbiased Estimator
Co-K	Co-Kriging
CoP	Code of Practice
CP	Competent Person
CPR	Competent Persons' Report
CRIRSCO	Committee for Mineral Reserves International Reporting Standards
CV	Competent Valuator
CW	Channel Width
CY	Calendar Year
DCF	Discounted Cash Flow
DDH	Diamond Drill Holes
IDW	Inverse Distance Weighting
KED	Kriging with External Drift
LoM	Life-of-Mine
MBF	Mining Block Factor
MCF	Mine Call Factor
MRM	Mineral Resource Management
NPV	Net Present Value
OK	Ordinary Kriging
PCF	Plant Call Factor
PTO	Planned Task Observation
QA/QC	Quality Assurance and Quality Control
RK	Regression Kriging
RoM	Run-of-Mine
SABS	South African Bureau of Standards
SAIMM	South African Institute of Mining and Metallurgy
SAMREC	South African Code for Reporting of Mineral Resources and Mineral Reserves
SAMVAL	South African Mineral Asset Valuation Committee
SANAS	South African National Accreditation System
SANS	South African National Standard
SCF	Shaft Call Factor
SG	Specific Gravity
SK	Simple Kriging
SMU	Selective Mining Unit or Smallest Mining Unit
Sox	Sarbanes-Oxley Act
SRD	Surface Rock Dump
StK	Stratified Kriging
SW	Stoping Width
TD	Tailings Dam, see TSF

TEMs	Technical Economic Models
TSF	Tailings Storage Facility
UK	Universal Kriging
UTO	Unplanned Task Observation
WACC	Weighted Average Cost of Capital

Units

'	minutes
%	percentage
cm	centimetre
cm.g/t	centimetre gram per tonne (measure of value)
dpa	days per annum
g	grams
g/t	gram per tonne (measure of grade)
ha	hectare
k	1,000 units
kg	kilogram
km	kilometre
koz	kilo (thousand) ounces
kt	kilo (thousand) metric tonnes
ktpa	kilo (thousand) tonnes per annum (year)
ktpm	kilo (thousand) tonnes per month
kV	kilo Volts (Volts × 1,000)
l/s	litres per second
m	a metre
m/s	metres per second
m^2	a square metre – measure of area
m^3	a cubic metre – measure of volume
Mbcm	millions – bulk cubic metres
Mlpd	mega litres per day
mm	a millimetre
Moz	a million troy ounces
mpa	metres per annum
Mt	a million metric tonnes
Mtpa	a million metric tonnes per annum
°	degrees
°C	degrees centigrade
oz	a fine troy ounce equalling 31.1034768 g
pa	per annum
R	South African rand
R/kg	South African rand per kilogram
R/t	South African rand per metric tonne
Rm	million South African rand
RoMt	Run-of-Mine tonne
t	metric tonne
t/m^3	density measured as metric tonnes per cubic metre
tpa	metric tonne per annum
tpd	metric tonne per day
tphr	metric tonne per hour

tpm	metric tonne per month
USD	United States dollar
USD/oz	United States dollar per troy ounce
US$m	million United States dollars
USc/t	United States cents per tonne
USD/lb	United States Dollars per pound
USD/t	United States Dollars per tonne
USDm	a million United States Dollars
ZAR	South African Rand
ZAR/kg	Rand per kilogram
ZAR/lb	Rand per pound
ZAR/t	Rand per tonne
ZARm	South African Rand million

Definitions

Blasthole – A drill hole in a mine that is filled with explosives to blast loose a quantity of rock.

Block width – The average width at which a block of ore is estimated to be mined.

Bulk mining – Any large-scale or mechanised method of mining involving mass mining and extraction of orebodies.

Capital expenditure – Specific project or ongoing expenditure: for replacing or purchasing additional equipment, materials, or infrastructure.

Carbon-in-leach (CIL) – The recovery process in which gold is first leached from gold ore pulp by cyanide and simultaneously adsorbed onto activated carbon granules in the same vessel. The loaded carbon is then separated from the pulp for subsequent gold removal by elution. The process is typically employed where there is a naturally occurring gold adsorbent in the ore.

Carbon-in-pulp (CIP) – The recovery process in which gold is first leached from gold ore pulp by cyanide and then adsorbed onto activated carbon granules in separate vessels. The loaded carbon is then separated from the pulp for subsequent gold removal by elution.

Cash costs – Direct mining costs, direct processing costs, direct general and administration costs, consulting fees, management fees, bullion transport, and refining charges.

Channel – Watercourse, also in this sense sedimentary material course.

Comminution – The process by which ore is reduced in size to liberate the desired mineral from the gangue material, in preparation for further processing.

Concentrate – A metal-rich product resulting from a mineral enrichment process such as gravity concentration or flotation, in which most of the desired mineral has been separated from the waste material in the ore.

Crosscut – A horizontal underground drive developed perpendicular to the strike direction of the stratigraphy.

Cut-off grade – The lowest grade of mineralised rock determines whether it is economical to recover its gold content by further concentration.

Decline – A surface or sub-surface excavation in the form of a tunnel, which is developed from the uppermost point downwards.

Depletion – An accounting device, recognising the consumption of an ore deposit, a mine's principal asset.

Development – Underground work carried out for the purpose of opening up a material deposit, including shaft sinking, crosscutting, drifting, and raising.

Diamond drill – A rotary type of rock drill that cuts a core of rock that is recovered in long cylindrical sections.

Dilution – Waste, which is unavoidably mined and/or trammed with ore.

Dip – Angle of inclination of a geological feature/rock from the horizontal.

Drill-hole – Method of sampling rock that has not yet been exposed.

Dyke – A tabular igneous intrusion that cuts across the bedding or foliation of the country rock.

Elution – The chemical process of desorbing gold from activated carbon.

Face – The end of a drift, crosscut, or stope at which work is taking place.

Facies – A rock unit defined by its composition, internal geometry, and formation environment.

Fault – The surface of a fracture along which movement has occurred.

Feasibility study – A comprehensive study undertaken to determine the economic feasibility of a project; the conclusion will determine if a production decision can be made and is used for financing arrangements.

Fire assay– The assaying of metallic ores by methods requiring the use of furnace heat.

Footwall – The underlying side of an orebody or slope.

Geostatistics – The branch of applied statistics used to model spatial/temporal data and discern patterns in geological phenomena, which by nature are subject to space-related and time-related variations. Geostatistics automates the correlation of sampled and unsampled data, enabling us to make generalisations about a data set, e.g., processes, events, and entities in a specific area. Geostatistics takes correlation between samples into account. Samples in a stope/on a bench are more likely to be similar if the samples are closer together than if they are further apart.

Gold equivalent – Gold plus silver, or another metal, expressed in equivalent ounces of gold, using a conversion ratio dependent on prevailing gold and silver (or other metal) prices.

Grade – The measure of concentration of gold within mineralised rock.

Hanging wall – The overlying side of an orebody or slope.

Haulage – A horizontal underground excavation, which is used to gain access to the orebody and to transport mined ore.

Head grade – The average grade of ore fed to a Mill and/or Plant.

Hedging – Taking a buy or sell position in the futures market, opposite to a position held in the cash spot market, to minimise the risk of financial loss from an adverse price change.

Hydrothermal – Process of injection of hot, aqueous, and generally mineral-rich solutions into existing geological and/or structural features

Indicated Mineral Resource – An 'Indicated Mineral Resource' is that part of a Mineral Resource for which tonnage, densities, shape, physical characteristics, grade, and mineral content can be estimated with a reasonable level of confidence. It is based on exploration, sampling, and testing information gathered through appropriate techniques from locations such as outcrops, trenches, pits, workings, and drill holes. The locations are too widely or inappropriately spaced to confirm geological and/or grade continuity but are spaced closely enough for continuity to be assumed.

Inferred Mineral Resource – An 'Inferred Mineral Resource' is that part of a Mineral Resource for which tonnage, grade, and mineral content can be estimated with a low level of confidence. It is inferred from geological evidence and assumed but not verified geological and/or grade continuity. It is based on information gathered through appropriate techniques from locations such as outcrops, trenches, pits, workings, and drill holes that may be limited or of uncertain quality and reliability.

ISO 14000 – International Standards for Organisations to implement sound environment management systems.

Kriging – An interpolation method that minimises the estimation error in the determination of a mineral resource.

Life of Mine (LoM) – Number of years that an operation is planning to mine and treat ore and is derived from the current mining plan.

Lock-up gold – Gold that is 'tied' up as temporary inventory within a processing plant, or sections thereof – typically milling circuits.

Material assets – The mine and its associated infrastructure.

Measured Mineral Resource – A 'Measured Mineral Resource' is that part of a Mineral Resource for which tonnage, densities, shape, physical characteristics, grade, and mineral content can be estimated with a high level of confidence. It is based on detailed and reliable exploration, sampling, and testing information gathered through appropriate techniques from locations such as outcrops, trenches, pits, workings, and drill holes. The locations are spaced closely enough to confirm geological and grade continuity.

Mill width – Calculated width expressing the relationship between the total reef area excavated and the total mill tonnes milled from underground sources, reported in centimetres.

Milling – A general term used to describe the process in which the ore is crushed, ground, and subjected to physical or chemical treatment to extract the valuable metals to a concentrate or finished product.

Mine Call Factor – The ratio expressed as a percentage which the specific product accounted for in 'recovery plus residue' bears to the corresponding product 'called for' by the mine's measuring and evaluation methods.

Mineral Reserve – The economically mineable material derived from a Measured and/or Indicated Mineral Resource. It is inclusive of diluting and contaminating materials and allows for losses that are expected to occur when the material is mined. Appropriate assessments to a minimum of a Pre-Feasibility Study for a project and a LoM Plan for an operation must have been completed, including consideration of, and modification by, realistically assumed mining, metallurgical, economic, marketing, legal, environmental, social, and governmental factors (the modifying factors). Such modifying factors must be disclosed.

Mineral Resource – A 'Mineral Resource' is a concentration [or occurrence] of material of economic interest in or on the Earth's crust in such form, quality, and quantity that there are reasonable and realistic prospects for eventual economic extraction. The location, quantity, grade, continuity, and other geological characteristics of a Mineral Resource are known, estimated from specific geological evidence and knowledge, or interpreted from a well-constrained and portrayed geological model. Mineral Resources are subdivided, in order of increasing confidence with respect to geoscientific evidence, into Inferred, Indicated, and Measured categories.

Mineralised – Rock in which minerals have been introduced to the point of a potential ore deposit.

Minimisation – Minimising the kriging prediction variance with calculus.

Normal fault – Fault in which the hanging wall moves downward in relation to the footwall, where the downthrow side is in the direction of the dip of the fault.

Nugget effect – A measure of the randomness of the grade distribution within a mineralised zone.

Ongoing capital – Capital estimates of a routine nature, which are necessary for sustaining operations such as replacement or additional equipment, materials, or infrastructure.

Ore Reserve – A 'Mineral/Ore Reserve' is the economically mineable material derived from a Measured and/or Indicated Mineral Resource. It is inclusive of diluting materials and allows for losses that may occur when the material is mined. Appropriate assessments, which may include feasibility studies, have been conducted, considering and modifying realistically assumed mining, metallurgical, economic, marketing, legal, environmental, social, and governmental factors. These assessments demonstrate at the time of reporting that extraction is reasonably justified. Mineral/ore reserves are subdivided in order of increasing confidence in probable mineral/ore reserves and proved mineral/ore reserves.

Economic breakeven – The average mining value at which it is estimated that ore can be mined without profit or loss.

Payshoot – Linear to sub-linear zone within a reef, for which gold grades or accumulations are predominantly above the cut-off grade and/or are generally of higher grade than the surrounds.

Percentiles – Describe the grade at which the percentage of data lies below. A tenth percentile grade has 10% of the sample data less than the value and 90% greater than the value. The percentiles are generated by sorting the data from the lowest to the highest grade and then selecting the sample values for which the corresponding percent of the data set lies below. A series of percentile grades provide an assessment of the relative grade distributions.

Pillar – Generally, an area left in situ to help support the excavations in an underground mine.

Plant Recovery Factor – The percentage ratio of the mass of the specific mineral product recovered from ore treated at the plant to its specific mineral content before treatment.

Probable Ore Reserve – A 'Probable Mineral/Ore Reserve' is the economically mineable material derived from a Measured and/or Indicated Mineral Resource. It is estimated with a lower level of confidence than a Proved Mineral/Ore Reserve. It is inclusive of diluting materials and allows for losses that may occur when the material is mined. Appropriate assessments, which may include feasibility studies, have been carried out, including consideration of and modification by, realistically assumed mining, metallurgical, economic, marketing, legal, environmental, social, and governmental factors. These assessments demonstrate at the time of reporting that extraction is reasonably justified.

Project capital – Capital expenditure, which is associated with specific projects of a non-routine nature.

Proved Ore Reserve – A 'Proved Mineral/Ore Reserve' is an economically mineable material derived from a measured mineral resource. It is estimated to have a high level of confidence. It is inclusive of diluting materials and allows for losses that may occur when the material is mined. Appropriate assessments, which may include feasibility studies, have been carried out, including consideration of and modification by realistically assumed mining, metallurgical, economic, marketing, legal, environmental, social, and governmental factors. These assessments demonstrate at the time of reporting that extraction is reasonably justified.

Random variables – A random variable is a property at a location (u_α) with multiple possible outcomes.

Random sampling – When every item in the population has an equal chance of being chosen. The selection of every item is independent of every other selection.

Regular sampling – When samples are taken at regular intervals (equally spaced).

Realisation – An outcome from a random variable or joint set of outcomes from a random function.

Reef – A precious metal bearing stratiform tabular orebody.

Regression – As a special case of weighted least-squares prediction in the generalised linear model with orthogonal projections in linear algebra.

Reverse fault – A fault that dips towards the block that has been relatively raised. One block of ground has been displaced above another block along the fault line.

Seismic – Earthquake or earth vibration including those that are artificially induced.

Shaft – An opening cut downwards from the surface for transporting personnel, equipment, supplies, ore, and waste.

Shear – A deformation resulting from stresses that cause contiguous parts of ore or rock to slide relative to each other in a direction parallel to their plane of contact.

Shortfall – Negative difference between the tonnage calculated by the surveyor as ore sent to plant versus that accounted for as delivered to the plant.

Sill – A tabular igneous intrusion that parallels the planar structure of the surrounding rock.

Spatial continuity – A measure of change in a property of interest over distance.

Statistics – A branch of science concerned with mathematical methods for collecting, organising, and interpreting data, as well as drawing conclusions and making reasonable decisions based on such analysis.

Stope – Underground void created by mining.

Stratigraphy – The science of rock strata.

Strike – The strike of an inclined plane is the direction of a horizontal line drawn on that plane. It is always perpendicular to the dip direction.

Stripping – The process of removing overburden, or waste rock, to expose ore.

Stripping ratio – The ratio of the amount of waste rock removed per tonne of ore mined.

Sub-vertical shaft – An opening, cut below the surface, vertically downwards from an established surface shaft.

Surface sources – Ore sources, usually dumps, tailings dams, and stockpiles, located at the surface.

The base case – The base case, as established as part of the Financial Models. I.e., On currently accepted assumptions of economic parameters, costs, and revenues.

Tonnage discrepancy – The difference between the tonnage hoisted as ore and accounted for by the plant measuring methods. Discrepancy is referred to as a shortfall when the calculated tonnage is less than the tonnage accounted for by the plant or an excess when the opposite occurs.

Tonne(s) – Metric tonne(s) = 1,000 kg.

Valuation date – The date the valuation is effective.

Variogram – A measure of the average variance between sample locations as a function of sample separation.

Winze – A steeply inclined shaft driven to connect one mine level with a lower level.

Wireframe – A digital surface constructed from vertices with connecting straight lines or curves.

Statistical and mathematical symbols

$\gamma(h)$ a measure of dissimilarity versus distance. It is a spatial variance between two data points separated by the distance, h

γ Gamma

C_0 Nugget effect – discontinuity in the variogram at distances less than the minimum data spacing. A ratio of nugget/sill is known as relative nugget effect (%)

C Total sill

C_1 Sill of the first semi-variogram structure

C_2 Sill of the second semi-variogram structure

C_3 Sill of the third semi-variogram structure

β Beta – additive constant applied to create a standardised statistical moment a three-parameter log-normal distribution.

$C(h)$ Covariance

$N(h)$ number of all data point pairs separated by the distance, h

h lag distance. Separation between two data points

u_a data point on 2D or 3D space at the location, α

$u_a + h$ data point separated from $u\alpha$ by the distance, h

$z(u_a)$ numerical value of data point, u_a

$z(u_a + h)$ numerical value of data point, $u_a + h$

σ^2 Sill or the sample total variance –Interpret spatial correlation relative to the sill, level of no correlation.

v volume of sample

V volume of a block

i, j, k a subscript or integer power

a the range of influence or distance scaling parameter in a model semi-variogram

Formula sheet

a. **Derivation of Distance Formula** by the Pythagoras theorem:

$$AB^2 = AC^2 + BC^2$$

$$d^2 = \left(x_2{}^2 - x_1{}^1\right)^2 + \left(y_2{}^2 - y_1{}^1\right)^2$$

Taking the square root on both sides,

$$d = \left[\left(x_2{}^2 - x_1{}^1\right)^2 + \left(y_2{}^2 - y_1{}^1\right)^2\right]^{0.5}$$

This is called the distance between two points formulae

For 3D distance, use: $d = \left[\left(x_2{}^2 - x_1{}^1\right)^2 + \left(y_2{}^2 - y_1{}^1\right)^2 + \left(z_2{}^2 - z_1{}^1\right)^2\right]^{0.5}$

b. **Measures of central tendency**

   i. Arithmetic mean is the sum of the sample values divided by the number of samples.

$$\text{Arithmetic mean} = \frac{\text{sum of values}}{\text{number of data points or observations}}$$

   ii. The median is the middle value when the samples are sorted in ascending order, e.g., sort samples in increasing order of grade and pick the middle sample.

   iii. The mode is the most typical sample value and is usually read as the peak of the histogram. When a histogram has two peaks, we describe it as 'bi-modal'.

   iv. Herbert Sichel is credited for providing empirical work to establish Sichel's factor which, when applied to geometric mean, provides unbiased Sichel's mean. Sichel's factor is the back-transform of half the variance of the log-transformed data. Sichel's mean (also called the log-estimated mean) is the unbiased mean when a data set is log-normally distributed. The assumption of log-normality is absolutely necessary for Sichel's mean to be meaningful. So, while calculating Sichel's mean is not difficult, it is only valid under the strict condition of a data set having a log-normal. The Sichel's mean is calculated as

$$\text{log estimated mean} = \exp^{(\log \text{mean})} \times \exp^{\left(\frac{\log \text{variance}}{2}\right)}$$

c. **Measures of variability**

   i. The range is the difference between the maximum and minimum sample value:

$$\text{Range} = \text{maximum value} - \text{minimum value}$$

ii.   When the data is sorted in grade order, the interquartile range is the difference between the top and bottom quarter values:

$$\text{Interquartile range} = \text{upper quartile} - \text{lower quartile}$$

iii. The variance is the typical difference between each sample value and the mean grade:

$$\text{variance } (\sigma^2) = \frac{\text{sum of (each ample value} - \text{mean})^2}{(\text{number of samples } p - 1)}$$

iv.   The standard deviation is the square root of the variance and brings the number back into a grade sense rather than a grade-squared sense:

$$\text{Standard deviation } (\sigma) = \sqrt{\sigma^2}$$

v.   The Coefficient of Variation (also called COV or CV) is the variability relative to the mean grade. The COV is useful for comparing the variability between data sets whose typical grades may be quite different

$$\text{Coefficient of Variation} = \frac{\text{Standard deviation } (\sigma)}{\text{Mean } (\mu)}$$

vi.   Calculation of 2D experimental variogram in Python or Excel:

DEFINE THROUGH A FUNCTION

```
def semi_variogram(X0, Y0, C0, X1, Y1, C1):
    change in X-coordinate (dx) = (X0 – X1)
    change in Y-coordinate (dy) = (Y0 – Y1)
    squared change in X-coordinate (dx²) = dx²
    squared change in Y-coordinate (dy²) = dy²
    distance between two points ( h ) = (dx² + dy² )0.5
    Variogram = (C0 – C1)²
    if dx = 0:
    dx = 0.001
    direction = math.atan(dy/dx)
    #The mathematical azimuth is measured counterclockwise from EW and not
clockwise from NS as the conventional
    azimuth i
    azimuth = (math.pi*0.5-direction) *180/math.pi
    Semi_variogram( γ(h) )= 0.5*Variogram
    return [dx, dy, dx², dy², h, Variogram, direction, azimuth, Semi_variogram]
```

d. **Double-structured spherical semi-variogram model**
   IF $h = 0$:

$$\gamma(h) = 0$$

IF $0 < h \leq a_1$:

$$\gamma(h) = c_0 + c_1 \left[\frac{1.5h}{a_1} - 0.5 \left(\frac{h}{a_1} \right)^3 \right] + c_2 \left[\frac{1.5h}{a_2} - 0.5 \left(\frac{h}{a_2} \right)^3 \right]$$

IF $a_1 < h \leq a_2$:

$$\gamma(h) = c_0 + c_1 + c_2 \left[\frac{1.5h}{a_2} - 0.5 \left(\frac{h}{a_2} \right)^3 \right]$$

IF $h > a_2$:

$$\gamma(h) = c_0 + c_1 + c_2$$

e. **Nugget %** – modelled log-normal variogram parameters need to be re-scaled to reflect the variability of the data. One way to do this is to rescale the nugget effect according to the log variance and then distribute the remaining variability according to the sills:

$$\text{Nugget } \% = \frac{(\log \sigma^2 - \text{nugget})}{\text{total sill}}$$

f. **Covariance** – The covariance is computed by averaging the product of the sample grades from two samples separated by a specified distance and then subtracting the product. This gives us:
   Covariance = [(grade of sample (1) × grade of sample (2)) + (grade of sample (1) × grade of sample (3)) + (grade of sample (2) × grade of sample (3))] ÷ 3 × [average (grade of sample (1), grade of sample (1), grade of sample (2)) × average (grade of sample (2), grade of sample (3), grade of sample (3))]

g. **Correlogram** is the covariance standardised by:

h. (Standard deviations of all the first sample grades) × (Standard deviations of all the second sample grades)
   For the example, this is the same as:
   Correlogram = [(grade of sample (1) × grade of sample (2)) + (grade of sample (1) × grade of sample (3)) + (grade of sample (2) × grade of sample (3))] ÷ 3 – [average (grade of sample (1), grade of sample (1) , grade of sample (2)) × average (grade of sample (2), grade of sample (3), grade of sample (3))]/ [(std dev) (grade of sample (1), grade of sample (1), grade of sample (2)) × std dev (grade of sample (2), grade of sample (3), grade of sample (3))]

i. **Covariogram** $C(h)$ and its relationship to the semi-variogram $\gamma(h)$:
j. **Covariance = Total Variance – Semi-variogram**

$$\text{Covariance} = \text{Nugget}(C_0) + \text{Sill}(C_i) - \text{Semi\_variogram}$$

$$C(h) = C - \gamma(h)$$

Total Sill = σ^2 = population variance
Covariance functions: v = sample support and V = block support
$C(v,v) = (C_0 + C_1 + C_2) - \gamma(v,v)$ covariance relationship between a sample and every other sample
$C(V,V) = (C_0 + C_1 + C) - \gamma(V,V)$ covariance relationship between all discretised points in the block
$C(v,V) = (C_0 + C_1 + C_2) - \gamma(v,V)$ covariance relationship between a sample and every discretised point in the block

k. **Convert discretised variogram (right-hand side) into a covariance matrix**

$$C(h) = \sigma^2_x - \gamma(h)$$

l. **Convert variogram (left-hand side) into a covariance matrix**

$$C(h) = IF\ (\gamma(h) = 0, 1, 1 - \gamma(h))$$

m. **Estimation Variance**

$$\text{Estimation Variance } (\sigma^2_E) = \bar{C}(A.A) + \bar{C}(z,z) - 2\bar{C}(z,A)$$

n. **Kriging Variance** is calculated as the weighted sum of the variograms between the sample and the block, plus the LaGrange multiplier less variability contained within the block. Historically, the kriging variance was used as a selling point for kriging since it provides a measure of confidence in the estimate. This measure is based on the sample configuration surrounding a block and the variogram.

$$\text{Kriging Variance } (\sigma^2_K) = \bar{C}(A.A) - \sum_{i=1}^{n} wi\bar{C}(zi, A) + \lambda$$

OR

Kriging variance = sum (variogram (point to block distance) × kriging weight) × average (variogram (between each and every discretisation point in block size)) + LaGrange multiplier

The kriging variance can be summarised as:

- sum of the point to block variability weighted by the kriging weights
- minus variability within a block
- plus LaGrange Multiplier

Let us take a closer look at each of these components:

Point to block variability: This is the average of the variogram values for each of the sample to discretisation points.

- Sum of the point to block variability weighted by the kriging weights: The contribution of each sample to variability (or confidence) in the block is summarised by adding up all the samples to block variabilities but weighting them according to the influence each sample has had on the estimate (i.e., the kriging weights).
- Variability within a block: This is the average of the variogram values based on the distances between each and every discretised point inside the block. For large blocks, this average will be large and, since it is subtracted in the formula, will mean a reduced kriging variance.
- LaGrange multiplier: The LaGrange multiplier increases when the data surrounding a block has a sub-optimal geometry (clustered, sparse, or resulting in an extrapolated estimate). Inadequate data configuration results in an increased LaGrange multiplier and, therefore, a higher kriging variance.

o. **Block Variance**

$$\sigma^2{}_B = \sigma^2 - \gamma(A, A)$$

where: $\sigma^2 = $ Total variance $= $ Sill of the variogram $= C$

p. **Kriging Slope of Regression** estimates the slope of the regression equation between the estimated and true block grades. When the slope is '1.0', the estimated high grades and estimated low grades correspond accurately to the respective true high and low grades.

$$\text{Slope } (b) = \frac{\text{Cov}\{Z_V, Z_V^*\}}{\sigma_{Z_V^*}^2} = \frac{\sum_{i=1}^{n} \lambda_i \bar{C}(v_i, V)}{\sum_{i=1}^{n} \sum_{j=1}^{n} \lambda_i \lambda_i C(v_i, v_j)}$$

OR

$$\text{Kriging Slop of Regression (SLOR)} = \frac{\sigma_B{}^2 - \sigma_K{}^2 + \lambda}{\sigma_B{}^2 - \sigma_K{}^2 + 2\lambda}$$

where:

$\sigma^2{}_B$ = Block Variance

$\sigma^2{}_K$ = Kriging Variance

λ = Lagrange multiplier. The LaGrange multiplier increases when the data surrounding a block has a sub-optimal geometry (clustered, sparse, or resulting in an extrapolated estimate). Inadequate data configuration results in an increased LaGrange multiplier and, therefore, a higher kriging variance.

q. **Kriging Efficiency** estimates the percent overlap expected between the estimated block histogram and the histogram of the true block grades. When a block has a kriging efficiency of 100%, we expect a perfect match between the estimated and true grade distributions. The kriging efficiency decreases as the data becomes more sparse, clustered, or blocks are extrapolated rather than interpolated. Sometimes, the kriging efficiency can even be negative, signalling extremely poor estimates:

$$\text{Kriging Efficiency} = \text{KE} = \frac{\sigma_B^2 - \sigma_K^2}{\sigma_B^2} \quad \text{or} \quad \frac{\text{Block Variance} - \text{Kriging Variance}}{\text{Block Variance}}$$

Kriging Efficiency in terms of the covariogram is given by:

$$\text{KE} = \frac{\bar{C}(V,V) - \sigma_K^2}{\bar{C}(V,V)}$$

where s_K^2 is the kriging variance and in terms of the covariogram is given by

$$\sigma_K^2 = \bar{C}(V,V) - \sum_{i=1}^{n} l_i \bar{C}(v_i, V) + \lambda$$

1 Setting the scene for mineral resource estimation

INTRODUCTION

> "It is impossible to know the true *in-situ* value of a panel, block, or bench before it is fully mined out. Therefore, specific techno-economic criteria must be met before a panel is selected for mining. Geostatistics provides tools for valuing blocks comprehensively and quantifying uncertainty. As blocks are mined out, it is necessary to validate mineral resource estimates through follow-up reconciliation studies." Prof. Danie G. Krige (1996).

Before we begin, it is important to understand the concept of ore, gangue, and mineral beneficiation, as these terms will be often used in this book. Ore is a naturally occurring material that contains one or more valuable minerals that potentially can be extracted at a profit. Ores typically occur in rock formations and can be mined to obtain the valuable minerals they contain (Abaka-Wood et al., 2022). Gangue refers to the waste rock or other material that is mixed with the ore during its formation and which has no commercial value. Gangue may also refer to the minerals or elements that are present in the ore but are not of economic value and must be removed during the beneficiation process to extract the desired minerals. Mineral beneficiation refers to the process of separating valuable minerals from their ores, with the aim of improving their economic value (Koruprolu and Nirmala, 2023). This can involve a range of techniques such as crushing, grinding, screening, washing, magnetic separation, flotation, and leaching (Quast et al., 2015). The mining industry is known as being a capital-intensive and high-risk business, with the only product being beneficiated ore(s) (Abaka-Wood et al., 2022). Investors and regulatory frameworks require that mining companies identify and quantify the uncertainty (and associated risk) that occurs throughout the mining value chain (Koruprolu and Nirmala, 2023). The mining value chain refers to the sequence of activities and processes involved in the exploration, extraction, processing, and marketing of minerals or other geological materials. It involves a range of activities, from the discovery of mineral deposits to the delivery of finished products to customers. The main components of the mining value chain are (Pietrobelli et al., 2018):

- Exploration: This involves the search for mineral deposits and the evaluation of their commercial viability.

DOI: 10.1201/9781032650388-1

1

- Mining: This involves the extraction of minerals from the ground through various methods such as underground mining, open-pit mining, or surface mining.
- Ore processing: This involves the physical and chemical treatment of the extracted minerals to remove impurities and produce a concentrate that can be sold to customers.
- Refining: This involves further processing of the concentrate to produce a final product that meets customer specifications (Goodfellow and Dimitrakopoulos, 2017).
- Marketing and sales: This involves the promotion and sale of the final product to customers, such as metal fabricators, electronics manufacturers, or construction companies.
- Environmental and social responsibility: This involves the management of environmental impacts and the implementation of social programmes to benefit local communities and stakeholders (Chipangamate et al., 2023).
- Research and development: This involves the continuous improvement of technologies and processes to optimise the mining value chain and to discover new mineral deposits.

RISK QUANTIFICATION AND GEOLOGICAL FACTORS

Risk quantification begins with identifying an exploration target and continues through the declaration of exploration results, mineral inventory estimation, and classification (Rose, 1987). Technical studies are then undertaken at various levels of detail to determine what portion of the mineral inventory can become a declared resource and be subsequently transformed into an economically viable mine. When a measured mineral resource or a proved mineral reserve is declared, it is implicitly assumed that the project's risk profile is lower than one where the mineralised material has been reported at the inferred level of confidence, indicating a higher degree of uncertainty and, thus a higher risk profile. It should be noted that a mineral reserve cannot be declared based on a mineral resource estimate classified at the Inferred level of confidence.

Several factors make each mineral deposit unique, such as the mineralising (ore paragenesis) processes, geographic location, socio-economic, and political settings, as well as environmental factors (Dominy et al., 2002). These factors are referred to as the 'modifying factors' by the Committee for Mineral Reserves International Reporting Standards (CRIRSCO) family of reporting codes. Each of the modifying factors carries its own risk weighting that may significantly affect the viability of a mining (or exploration) project. Even if social, economic, political, and environmental concerns are ignored, the characteristics of the mineralised deposit, in terms of contained tonnage, metal grade (or deleterious elements/contaminants), and geological continuity, need to be understood. Additionally, the quality and distribution of the mineral or metal contained within the mineralised deposit must also be understood.

Physico-chemical and metallurgical properties of the deposit affect the whole mining value chain. This makes it crucial to analyse the uncertainties and variabilities that occur at the various stages of the mining value chain using a systems approach that places mine-to-mill reconciliation processes at the centre of the risk assessment process. Estimating recoverable mineral resources is a dynamic process required for operational resiliency, governance, reporting, and risk quantification. In view of this, it is essential to identify and evaluate all potential errors and uncertainties associated with the mineral estimation processes at an early stage. In general, there are four areas of uncertainty in the mineral resource estimation process: (a) the inherent variability that occurs in a mineral deposit; (b) data gathering and assay errors; (c) errors at different stages of the resource estimation process due to poor practice, incorrect geological interpretations due to limited geological understanding; and/or (d) the application of inappropriate geostatistical methodologies.

Natural variability is what makes each deposit unique. Different deposits vary in terms of their mineral genesis and paragenesis, deposit geology and deposit morphology, mineral assemblage (mineralogy), mineral size distribution, degree of oxidisation and the presence of deleterious elements. Of particular interest in this book are stratabound and stratiform mineral deposits. Stratabound mineral deposits are confined to a single stratigraphic unit (i.e. they can also include variously orientated deposits or veins within a unit) and examples include: the great zinc Pine Point deposit in northern Canada, as well as the zinc-lead-silver Mt. Isa-McArthur deposit in Australia. Stratiform mineral deposits are a special type of stratabound mineral deposits where the deposits are strictly co-extensive with the stratigraphic unit. Examples include the platinum-chromite layers of the Bushveld Igneous Complex and the Witwatersrand Basin gold deposits, both found in South Africa. Depending on how the variability in the deposit is defined mathematically, it can be divided into structured and unstructured variability. Structured variabilities, such as the compositional variation seen in layered igneous intrusions (e.g. Bushveld Igneous Complex in South Africa) or cyclic depositional patterns in sedimentary rocks (e.g. Witwatersrand Basin in South Africa), can be explained mathematically; however, unstructured variability is erratic and unpredictable and thus cannot be defined mathematically (Rupprecht and Njowa, 2016). Uncertainty due to inadequate controls is influenced by poor practices, limited knowledge, and different biases that occur at and in each stage of a mineral resource estimation (Githiria and Musingwini, 2019). Errors are typically incurred during:

a. Geological data collection: This refers to the process of collecting data related to the geology of the deposit, which includes activities such as surveying, mapping, and various exploration techniques such as pitting, trenching, and drilling. Errors can occur during these data collection activities, leading to inaccurate information.

b. Geotechnical data collection: Geotechnical data collection involves gathering information about the physical and mechanical properties of the rocks and soils in the deposit. This data is essential for assessing the stability and suitability of the mining operations. Errors in geotechnical data collection can result in an inadequate understanding of the deposit's geotechnical characteristics, leading to potential risks and uncertainties.

c. Sampling, sub-sampling, and assaying involve selecting representative samples from the deposit, dividing them into smaller sub-samples, and analysing them to determine their mineral content, bulk chemistry, and/or grade. Errors in sampling, sub-sampling, or assaying can introduce biases and inaccuracies in the estimation of mineral resources.

d. Bulk density and mass determination: Bulk density refers to the mass of a unit volume of material, and it is an important parameter for calculating the tonnage of mineral resources. Errors in determining the bulk density or mass of the material can lead to incorrect estimations of the resource quantities.

e. *In-situ* moisture content: In certain deposits, such as coal, the moisture content of the material can significantly affect its quality and quantity. Errors in determining the *in-situ* moisture content can impact the estimation of coal resources.

f. Calibration of measuring instruments: Instruments used for measuring various properties, such as radiometric probes for detecting radiation levels, need to be calibrated accurately. Errors in the calibration process can result in incorrect measurements and subsequent erroneous resource estimations.

g. Quality assurance and quality control (QA/QC) procedures: QA/QC procedures are implemented to ensure the accuracy, precision, and reliability of data throughout the estimation process. Errors in implementing proper QA/QC protocols can introduce uncertainties and compromise the validity of the resource estimates.

h. Geological interpretation and modelling: Geological interpretation involves analysing the collected data and constructing a geological model of the deposit. Errors or biases in interpreting the geological information or in the modelling process can lead to incorrect representations of the deposit's characteristics, affecting the resource estimation.

i. Grade and tonnage estimation: Estimating the grade (concentration of valuable minerals) and tonnage (quantity of the material) is a crucial step in mineral resource estimation. Errors in the estimation methods or calculations can result in significant discrepancies in the resource estimates.

j. Mineral resource estimate validation: Validating the accuracy and reliability of the mineral resource estimate involves comparing the estimation results with independent data or using statistical methods. Errors in the validation process can lead to misleading or unreliable resource estimates.

k. Mineral resource classification and reporting: Mineral resource classification involves categorising the estimated resources based on their level of confidence and geological certainty. Errors in the classification and reporting of resources can result in misrepresentation or improper communication of the deposit's potential value and risks.

This book aims to provide guidance on current practices in mineral resource estimation, mineral reserve conversion and mine-to-mill reconciliation to assist in reducing technical risks to exploration and mining projects and promote transparency in the reporting of mineral resources and mineral reserve estimates (see Figure 1.1).

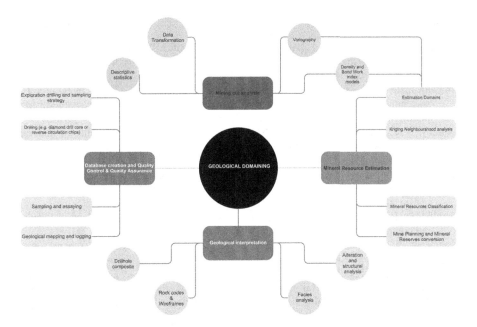

FIGURE 1.1   Simplified process workflow for mineral resources estimation.

CONCLUSION

In this chapter, we set the stage for understanding the intricate world of mineral resource estimation and the mining value chain. The fundamental challenge lies in assessing the true *in situ* value of mineral deposits before they are fully mined out. To address this challenge, specific techno-economic criteria must be met before selecting a panel for mining, and geostatistics provides valuable tools for comprehensive block valuation and uncertainty quantification. Mining, being a capital-intensive and high-risk business, necessitates identifying and quantifying uncertainties and risks throughout the mining value chain. This chain comprises various components, including exploration, mining, ore processing, refining, marketing, environmental and social responsibility, and research and development. Throughout the mining value chain, errors can occur at several stages, including geological and geotechnical data collection, sampling, sub-sampling and assaying, bulk density and mass determination, in-situ moisture content assessment, calibration of measuring instruments, Quality Assurance and Quality Control (QA/QC) procedures, geological interpretation and modelling, grade and tonnage estimation, mineral resource estimate validation, and mineral resource classification and reporting. Understanding the intricacies of mineral resource estimation and risk assessment is key for the successful and sustainable operation of mining projects. Through the application of geostatistics and systematic risk analysis, mining companies can make informed decisions, optimise their mining value chain, and ensure responsible resource management for the benefit of all stakeholders.

REFERENCES

Abaka-Wood, G. B., Johnson, B., Addai-Mensah, J., & Skinner, W. (2022). Recovery of Rare Earth Elements minerals in complex low-grade saprolite ore by froth flotation. *Minerals*, *12*(9), 1138.

Chipangamate, N. S., Nwaila, G. T., Bourdeau, J. E., & Zhang, S. E. (2023). Integration of stakeholder engagement practices in pursuit of social licence to operate in a modernising mining industry. *Resources Policy*, *85*, 103851.

Dominy, S. C., Noppé, M. A., & Annels, A. E. (2002). Errors and uncertainty in mineral resource and ore reserve estimation: The importance of getting it right. *Exploration and Mining Geology*, *11*(1–4), 77–98.

Githiria, J., & Musingwini, C. (2019). A stochastic cut-off grade optimization model to incorporate uncertainty for improved project value. *Journal of the Southern African Institute of Mining and Metallurgy*, *119*(3), 217–228.

Goodfellow, R., & Dimitrakopoulos, R. (2017). Simultaneous stochastic optimization of mining complexes and mineral value chains. *Mathematical Geosciences*, *49*, 341–360.

Koruprolu, V. B. R., & Nirmala, G. (2023). Development of process flow scheme on a low-grade calcareous graphite ore by split flotation technique. *Separation Science and Technology*, *58*(3), 509–519.

Krige, D. G. (1996). A practical analysis of the effects of spatial structure and data available and used, on conditional biases in ordinary kriging – 5th International Geostatistics Congress, Wollongong, Australia, 1996. (Vol. 7.16).

Pietrobelli, C., Marin, A., & Olivari, J. (2018). Innovation in mining value chains: New evidence from Latin America. *Resources Policy*, *58*, 1–10.

Quast, K., Connor, J. N., Skinner, W., Robinson, D. J., & Addai-Mensah, J. (2015). Preconcentration strategies in the processing of nickel laterite ores Part 1: Literature review. *Minerals Engineering*, *79*, 261–268.

Rose, P. R. (1987). Dealing with risk and uncertainty in exploration: How can we improve? *AAPG Bulletin*, *71*(1), 1–16.

Rupprecht, S., & Njowa, G. (2016, May). The valuation of an exploration project having Inferred Resources. In: *SAMREC SAMVAL Code Companion Volume Conference* (pp. 17–18). https://ujcontent.uj.ac.za/esploro/outputs/9913704807691

2 A brief history of mining geostatistics

INTRODUCTION

For both academics and industry professionals, understanding the history of geostatistics is crucial. The development of spatial analysis techniques and their use in diverse areas can be achieved by understanding the historical trajectory of geostatistics. Modern spatial analysis has been greatly affected by geostatistics, a field that integrates statistical analysis with spatial data. Researchers get a clearer understanding of the origins and evolution of important concepts, procedures, and tools that serve as the basis of contemporary spatial analysis by exploring the historical context of geostatistics. Examining the history of geostatistics sheds light on the motivations, challenges, and innovations that shaped the discipline. It allows researchers to understand the intellectual contributions of key figures such as Danie G. Krige and Georges Matheron, who laid the groundwork for geostatistics and introduced concepts such as variogram analysis and Kriging techniques. Additionally, knowing the background of geostatistics helps practitioners appreciate the importance and potential drawbacks of various methodologies. Practitioners can gain an understanding of the advantages, constraints, and suitable applications of various methodologies by studying the historical development of geostatistics. The history of geostatistics also offers a larger perspective on how technical improvements have affected spatial analysis. The discipline has seen a revolution with the introduction of computer technology, geostatistical software packages and artificial intelligence (includes machine learning). This has resulted in more advanced modelling and prediction capabilities.

GENESIS OF GEOSTATISTICS

"It can be expected that the gold values in a whole mine will be subject to a larger relative variation than those in a portion of the mine". Prof Danie G. Krige (1951).

The statement by Danie G. Krige (1951) suggests that samples taken close to each other are more likely to have similar values than if taken farther apart. This has become known as Waldo Tobler's (1970) First Law of Geography: "everything is related to everything else, but near things are more related than distant things".

Why did geostatistics emerge as a new discipline?

As mining progressed, the general practice has been, and still is, to regularly sample advancing stope or bench faces inside each ore block and compare these results with the original block estimation via the so-called block factors. Records kept over all the decades up to the end of the 1940s showed a disturbing feature that could not be explained, and thus, nothing was done about it:

- Blocks valued as low-grade were under-valued, and blocks valued as high-grade were over-valued.
- These conditional biases could cover a wide range, often exceeding −50% to +50%.

These biases were first explained by Krige (1951) using the lognormal distribution model, and standard correlation and regression techniques referred to as 'Elementary Kriging'. Regression factors were then introduced to eliminate conditional biases, and thus, to correct and improve block estimates. These were either determined theoretically or based on observed factors over recent years. Because such a regressed estimate is a weighted average of the peripheral grade and the population mean (i.e. the average of all data outside the block peripheries), these estimates formed the first application of what later came to be called 'Kriging'. It was in fact 'Elementary Kriging'. The correlation model used also introduced the concept of different types of data and distributions, that is, point or sample data and block grades or distributions – called 'change of support' in geostatistics.

Following a study of the spatial patterns evident in the data (i.e. correlation levels reducing as data lags for grade pairs increase and showing higher levels in preferred directions), Elementary Kriging soon developed into Simple Kriging with the weight for the population mean supplemented by weights for individual sample data close to the block. This improved the error variance of the estimates. Simple Kriging is, in effect, a multiple correlation model with the covariance matrices determined by the spatial structure as defined. To cover the possibility of trends in the data, with the population means being locally non-stationary, Ordinary Kriging was developed by replacing the population mean entirely with individual data in the area surrounding the block, with the Lagrange multipliers introduced for the matrix inversion. In mathematical optimisation, the method of Lagrange multipliers is a strategy for finding the local maxima and minima of a function subject to equation constraints (i.e. subject to the condition that one or more equations must be satisfied exactly by the chosen values of the variables). It is named after the French mathematician Joseph-Louis Lagrange.

The basic idea is to convert a constrained problem into a form such that the derivative test of an unconstrained problem can still be applied. The relationship between the gradient of the function and gradients of the constraints rather naturally leads to a reformulation of the original problem, known as the Lagrange multiplier method. The danger here is that a smoothing bias can still be present. If too much data are accessed for a block estimation, the block's estimated grade will approach the mean of the samples. Furthermore, if insufficient data are accessed, one obtains a conditional bias (mentioned earlier) where the low grades are underestimated, and high grades are overestimated. At a high enough amount of data accessed, estimates will approach the population mean; the Lagrange multiplier will then tend to become insignificant, and the Ordinary Kriging estimate will tend to approach the Simple Kriging version, that is, the equivalent of a classical statistical estimate – and will be conditionally unbiased.

THE EVOLUTION JOURNEY OF GEOSTATISTICS

Geostatistics has its roots in applied statistics, statistical mechanics, and calculus. Some of the technical breakthroughs date back to Kolmogorov and Wiener's independent work on time series in the 1930s and 1940s (Journel, 1983). At the beginning of the 1950s, estimation of regionalised variables, that is, the estimation of the value of a point in space, taking into consideration spatial trends associated with the characteristic that is being estimated such as density, thickness, or metal concentration, was simply the arithmetic average of the data belonging to the area that was being estimated, referred to by practitioners as a 'panel'. The estimation methodology was relatively simplistic in that only the samples located on or around its border were used.

During the 1950s and continuing into the 1960s, the discipline of geostatistics began to evolve at the same time as the disciplines of mining, meteorology, and forestry. This was led by the discoveries made by Daniel (Danie) Gerhardus Krige (26 August 1919 to 2 March 2013), a mining engineering graduate from the University of the Witwatersrand, South Africa. Through studying mineral resource estimation data from the Witwatersrand goldfields (South Africa), Krige (1951, 1962) observed that panels selected based on a high cut-off grade were, on average, less economical than expected when they were mined and sampled, that is, the estimated grade was higher than the actual sampled grade.

The most constant theoretical development of the theory of regionalised variables occurred at the School of Mines in Paris when Georges François Paul Marie Matheron (2 December 1930 to 7 August 2000) and co-founder, Jean-Paul Frédéric Serra (1940 to present), became interested in Krige's work. Krige noticed that the grades of mining panels or stopes were less variable than the grades of the chip or core samples used to estimate them, and he had developed a regression technique to account for this effect (Krige, 1951). Matheron incorporated Krige's findings into a general-purpose spatial estimating tool with a built-in support effect. In honour of Danie G. Krige's pioneering work, Matheron named this method 'Kriging'. According to Cressie (1990), the originally French term 'Krigeage' was coined by Pierre Carlier and first used at the French Commissariat à l'Energie Atomique in the late 1950s. Subsequently, 'Krigeage' was translated to 'Kriging' by Matheron (1963b). 'Krigeage' first appeared in Matheron's work in 1960, where it is referred to as a concept already known at the time. Matheron (1962, 1963a) presented the theory of Kriging as it is commonly understood.

Georges Matheron (1962) defined geostatistics as the application of probabilistic methods to regionalised variables, which designates any function distributed in real space.

Drawing inspiration from this conceptualisation, Matheron went on to found the famed Centre of Geostatistics and Mathematical Morphology at the École des Mines (Centre de Géostatistique et de Morphologie Mathématique) in Fontainebleau (Paris, France) in 1968, where his early students went on to achieve great things in geostatistics, morphology, and pure mathematics. Mathematical Morphology is a powerful technique that has been developed for analysing and processing a wide range of geometrical structures, including digital images, graphs, surface meshes, solids, and other spatial structures (Matheron, 1967). It is based on the principles of set theory, lattice theory, topology, and random functions. It has been extensively used in various disciplines, such as computer vision, image processing, geostatistics, and remote sensing.

Geostatistics is one of the disciplines where Mathematical Morphology finds significant applications. It is a discipline that deals with analysing and interpreting spatial data. In geostatistics, Mathematical Morphology is used to extract valuable information from the data by analysing its geometrical structures. For example, it can be used to extract the edges of geological features, such as faults and fractures, from digital images of the Earth's surface. Another example of the use of Mathematical Morphology in geostatistics is the analysis of spatial patterns of minerals in mineral deposits. Through analysing the geometrical structures of mineral grains in a rock sample, it is possible to extract information about the grains' size, shape, and orientation. This information can then be used to estimate the quality and quantity of the mineral deposit. Mathematical Morphology was first introduced in 1964 at the same institute, and since then, it has become an important tool for analysing and processing geometrical structures. Its versatility and wide range of applications make it a valuable technique in various disciplines, including geostatistics.

Various other practitioners contributed to developing and advancing this new, applied discipline of applied statistics, including such luminaries as Andre Journel, Clayton V Deutsch, Christian Lantuéjoul, Christina Dohm, Edward H. Isaaks, Isobel Clark, Jacqui Coombes, Jacques Rivoirard, Jame Gomez-Hernandez, Jef Caers, Jean-Michel M. Rendu, Jean Serra, Jean-Paul Chiles, Julián Ortiz, Margaret Armstrong, Marat Abzalov, Michel David, Michael J. Pyrcz, Mohan Srivastava, Nasser Madani, Oy Leuangthong, Peter A. Dowd, Pierre Delfiner, Pierre Goovaerts, Roussos Dimitrakopoulos, Winfred Assibey-Bonsu, and Xavier Emery. Significant theoretical breakthroughs occurred in the 1970s and included the measurement of uncertainty in Kriged estimates (Delfiner, 1999). This led to Kriging being applied in other disciplines such as cartography (Matheron and Huijbregts, 1971), oceanography (Chilès and Chauvet, 1975), meteorology (Chauvet et al., 1976; Delfiner and Delhomme, 1973), hydrology (Delhome, 1978), design of aircraft (Chung and Alonso, 2002), the prediction of the mechanical properties of nanomaterials (Yan et al., 2012), the optimisation of supply chain networks (Dixit et al., 2016), the construction of financial term-structures (Cousin et al., 2016), and the modelling of social systems (De Oliveira et al., 2013).

MATURITY

A general slowdown in the discipline's theoretical advances occurred as the subject matured between the 1980s and the 1990s. However, the use and application of Krige's methodology spread into the mining industry, and by the 2000s, it had become the *de facto* method to estimate mineral resources. Further development of Danie's M.Sc. thesis led to the development of various sub-techniques, such as simulation and conditioning, as well as a form of Gaussian Process Regression integrated into machine learning algorithms. One of the earliest and most widely used simulation techniques in geostatistics is known as unconditional simulation. In this method, data are simulated without any constraints or conditioning, using parameters estimated from the entire data set. This technique is useful for generating alternative realisations of the underlying spatial structure that may be consistent with the available data. Conditional simulation, on the other hand, is a simulation technique that uses data from a specific location to condition the simulation at that location. This method is used to generate realisations of the spatial structure that are consistent with the data at the conditioning location while preserving the spatial correlation and variability of the data in the surrounding areas.

Another simulation technique that has become increasingly popular in geostatistics is the Plurigaussian Simulation method. In this method, different geological facies are modelled separately, each with its own statistical distribution and spatial correlation. The approach allows for more realistic and accurate modelling of geological heterogeneity in the subsurface. Indicator Kriging is a type of geostatistical simulation that deals with binary data or categorical variables. In this method, each category or binary variable is assigned a unique number or indicator value, and the spatial correlation between these indicator values is estimated and used to generate realisations of the subsurface.

Multiple-point simulation is another technique used in geostatistics, which takes into account multiple-point statistics, such as the distribution and arrangement of local patterns and structures. This method is useful in modelling complex subsurface structures that traditional geostatistical methods cannot capture. Uniform conditioning is a technique used to estimate a variable's distribution at a specific location, given information about the variable at other locations. The technique is used when the data are sparse or when there are irregularities in the data. It assumes that the variable being estimated has a constant mean and variance over the area being considered. Localised uniform conditioning is a modification of uniform conditioning, which allows the estimation of the variable's distribution in a smaller region around the location being considered (Abzalov, 2006). Typically, the technique is useful when there are changes in the variable's mean or variance over the area being considered. Direct conditioning is a technique that involves the use of hard data or secondary information to condition the estimation of a variable's distribution. In this method, the estimate of a variable's distribution is modified based on the available data or information.

The evolution of geostatistics owes much to the development of innovative techniques that have contributed to its practical applications across a wide range of industries. These techniques have played a crucial role in estimating mineral reserves, predicting the mechanical properties of nanomaterials, optimising supply chain networks, and modelling complex social systems. With so many exciting opportunities in geostatistics, there has never been a better time to learn its fundamental principles and contribute towards its continued advancement. So why not embark on this fascinating journey of discovery and unlock the full potential of geostatistics?

GEOSTATISTICS IN THE DIGITAL ERA

The digital era has significantly transformed geostatistics, opening new avenues for spatial analysis and expanding its applications across diverse disciplines. This section explores the intersection of geostatistics with the digital era, focusing on its applications in geomatics, integration with artificial intelligence, and the role of geostatistical software packages in democratising spatial analysis. Geostatistical techniques have found extensive application in remote sensing and satellite imagery analysis. Researchers can extract valuable information about the Earth's surface and dynamics by combining geostatistical methods with remotely sensed data. Geostatistical interpolation techniques, such as Kriging, are used to generate high-resolution maps by assimilating remotely sensed observations. This integration enables the estimation of spatial patterns, monitoring of land cover changes, and identifying environmental phenomena at various scales. The integration of geostatistics with artificial intelligence has further expanded the capabilities of spatial analysis.

Machine learning algorithms, a subset of artificial intelligence, such as random forests and support vector machines, have been combined with geostatistical approaches to improve prediction accuracy and handle complex spatial relationships. Fusing these techniques enables the integration of diverse data sources, including geospatial and non-geospatial data, to extract meaningful insights and make accurate predictions. Geostatistical software packages have played a crucial role in democratising spatial analysis, making it more accessible to a broader audience. These software packages provide user-friendly interfaces and a wide range of geostatistical tools, enabling researchers, practitioners, citizens, and students to perform spatial analysis without extensive programming knowledge. Users can visualise data, apply geostatistical techniques, and generate maps and spatial models with relative ease. This accessibility has led to increased adoption of geostatistics across industries, such as environmental management, urban planning, and natural resource exploration.

CONTEMPORARY CHALLENGES AND FUTURE DIRECTIONS

The field of geostatistics continues to evolve, presenting both contemporary challenges and exciting future directions. This section highlights some of the current challenges in geostatistical modelling and analysis, explores new frontiers in geostatistics, and discusses potential applications and developments on the horizon. Geostatistical modelling and analysis face several challenges in the contemporary landscape. One key challenge is dealing with 'big data' and handling the increasing volume, velocity, and variety of spatial data. Geostatisticians must develop innovative approaches to process and analyse large datasets efficiently, ensuring scalability and computational efficiency. Another challenge lies in modelling complex spatial relationships and incorporating additional sources of uncertainty. Spatial phenomena often exhibit non-linear patterns, anisotropy, and multiscale variations, requiring advanced modelling techniques that go beyond traditional linear geostatistical methods. Additionally, incorporating multiple sources of uncertainty, such as spatial-temporal variability and measurement errors, is a challenge that geostatisticians must address to improve the accuracy of predictions and decision-making.

As geostatistics advances, new frontiers are emerging that push the field's boundaries. One frontier is the integration of geostatistics with other disciplines, such as ecology, epidemiology, and social sciences. By combining geostatistics with these domains, researchers can explore complex spatial interactions, understand ecological processes, model disease spread, and analyse spatial patterns in social phenomena. The integration of geostatistics with emerging technologies, such as the Industrial Internet of Things and sensor networks, opens exciting possibilities. Real-time geostatistical modelling, leveraging continuous streams of data from sensors and Internet of Things devices, can provide valuable insights for monitoring environmental conditions, optimising resource allocation, and improving decision-making in various domains. Looking ahead, geostatistics holds great potential for applications in fields such as precision agriculture, urban planning, and climate change adaptation. Geostatistical techniques can help optimise resource allocation in agriculture, model urban growth and transportation patterns, and assess the impacts of climate change on spatially distributed phenomena.

CONCLUSION

In conclusion, contemporary challenges in geostatistical modelling include handling big data, modelling complex spatial relationships, and incorporating additional sources of uncertainty. However, the future of geostatistics is promising, with new frontiers emerging through interdisciplinary collaborations and integration with emerging technologies. The potential applications of geostatistics in disciplines such as agriculture, urban planning, and climate change adaptation highlight its ongoing relevance and impact on addressing real-world challenges and advancing our understanding of spatial phenomena.

REFERENCES

Abzalov, M. Z. (2006). Localised uniform conditioning (LUC): A new approach for direct modelling of small blocks. *Mathematical Geology*, *38*, 393–411.

Chauvet, P., Pailleux, J., & Chilès, J. P. (1976). Analyse objective des champs météorologiques par co-krigeage. *La Météorologie*, *6*(4), 37–54.

Chilès, J. P. & Chauvet, P. (1975). Kriging: A method for cartography of the sea floor. *The International Hydrographic Review*, *52*(1), 25–41.

Chung, H. S. & Alonso, J. (2002, January). Using gradients to construct cokriging approximation models for high-dimensional design optimization problems. In *40th AIAA aerospace sciences meeting & exhibit* (p. 317). https://arc.aiaa.org/doi/10.2514/6.2002-317

Cousin, A., Maatouk, H., & Rullière, D. (2016). Kriging of financial term-structures. *European Journal of Operational Research*, *255*(2), 631–648.

Cressie, N. (1990). The origins of kriging. *Mathematical Geology*, *22*, 239–252.

De Oliveira, M. A., Possamai, O., Dalla Valentina, L. V. & Flesch, C. A. (2013). Modeling the leadership–project performance relation: Radial basis function, Gaussian, and Kriging methods as alternatives to linear regression. *Expert Systems with Applications*, *40*(1), 272–280.

Delfiner, P. (1999). *Geostatistics: Modelling spatial uncertainty*. New York: Wiley.

Delfiner, P. & Delhomme, J. P. (1973). *Optimum interpolation by kriging: Proceedings of NATO-ASI*. Nottingham: Display and Analysis of Spatial Data.

Delhome, J. P. (1978). Kriging in the hydrosciences. *Advances in Water Resources, 1*, 251–266.

Dixit, V., Seshadrinath, N., & Tiwari, M. K. (2016). Performance measures-based optimization of supply chain network resilience: A NSGA-II+ Co-Kriging approach. *Computers & Industrial Engineering, 93*, 205–214.

Journel, A. G. (1983). Nonparametric estimation of spatial distributions. *Mathematical Geology, 15*(3), 445–468.

Krige, D. G. (1951). A statistical approach to some basic mine valuation problems on the Witwatersrand. *Journal of the Chemical, Metallurgical and Mining Society of South Africa, 52*, 119–139.

Krige, D. G. (1962). Effective pay limits for selective mining. *Journal of the South Africa Institute of Mining and Metallurgy, 62*, 345–363.

Matheron, G. (1962). *Traité de géostatistique appliqué, vol. I: Mémoires du Bureau de Recherches Géologiques et Minières, no. 14* (333 pp.). Paris: Éditions Technip.

Matheron, G. (1963a). *Traité de géostatistique appliqué, vol. II, Le krigeage: Mémoires du Bureau de Recherches Géologiques et Minières, no. 24* (171 pp.) Paris: Éditions Bureau de Recherche Géologiques et Minières.

Matheron, G. (1963b). Principles of geostatistics. *Economic Geology, 58*, 1246–1266.

Matheron, G. (1967). Kriging or polynomial interpolation procedures? *CIMM Transactions, 70*, 240–244.

Matheron, G. & Huijbregts, C. (1971). Universal Kriging (an optimal method for estimating and contouring in trend surface analysis), decision making in the mineral industry. *CIM Special Volume 12*, 159–169.

Tobler, W. (1970). A computer movie simulating urban growth in the Detroit region. *Economic Geography, 46*(2), 234–240.

Yan, S., Shi, Y., Xiao, Z., Zhou, M., Yan, W., Shen, H., & Hu, D. (2012). Development of biosensors based on the one-dimensional semiconductor nanomaterials. *Journal of Nanoscience and Nanotechnology, 12*(9), 6873–6879.

3 Sampling for mineral resources estimation

INTRODUCTION

Sampling has been an essential practice for thousands of years, with its origins dating back to ancient Greek and Egyptian societies. In ancient times, sampling was commonly used for sampling vineyards for wine production, with random sampling using 'lots' explicitly mentioned on several occasions in the Holy Bible. However, the rigorous mathematical treatment of sampling only began with the introduction of probability theory in the 16th century by Pierre de Fermat and Blaise Pascal (Weil, 1973). Later, in 1812, Pierre Simon de Laplace combined probability theory with calculus in his treatise *Théorie Analytique des Probabilitiées*, providing a framework for future statistical analysis (Hahn, 2005).

In this chapter, our focus is not on the statistical analysis of sampling data but rather on the fundamental concepts that underlie the sampling process. These concepts are based on reason and logic, and can be categorised as problem definition, relevance, methods, truth, causality, the influence of time and space, reduction of bias, representivity, accuracy and precision, and types of sampling.

PROBLEM DEFINITION

In the context of mineral resources sampling, problem definition is the first step in the process of accurately estimating the amount and quality of minerals in a deposit. It involves identifying the objectives of the study, the variables that need to be measured, and the resources available for the project. The project manager should be clear about the intended outcome and the information required to achieve it. This includes identifying the type of mineral deposit, its location, and the size of the area to be sampled.

The problem definition should also consider the sampling method to be used. There are various methods of sampling available, including random, systematic, stratified, and cluster sampling. The choice of sampling method will depend on the nature of the mineral deposit, the objectives of the study, and the resources available to complete the project.

Moreover, problem definition should consider the time frame for project completion. The duration of the sampling exercise will depend on the size of the area to be sampled, the sampling method used, and resources available. The project manager should ensure that there is enough time to collect sufficient data, while adhering to the project's timeline. It is also essential to understand the relationship between the variables that need to be measured. For example, mineral deposits may be associated with specific geological formations or structures, and it may be necessary to collect data on these variables to

DOI: 10.1201/9781032650388-3

accurately estimate the mineral resources. The project manager should ensure that the sampling plan is designed to capture these relationships between variables accurately.

RELEVANCE

There may be times when one cannot directly sample the material of interest. Researchers need to ensure that they are sampling variables relevant to the required outcome. Consider the case of an apple orchard. Although there is a correlation between the amount of rain and the subsequent quantity and quality of the apples obtained, this would certainly not be a valid measure of when harvesting should occur. Here, the researcher is more concerned about the ripeness of the apples, and as such, the apples themselves need to be sampled and not the rainfall for the period, as this is not a relevant factor. For example, consider a gold deposit where the grade of the ore is the key factor in determining its economic viability. In this case, it would not be relevant to sample the thickness of the overburden or the colour of the rock. Instead, the focus should be on collecting samples of the ore itself and measuring its grade.

It is important to note that relevance may not always be immediately obvious, and some forethought is necessary to ensure that the right variables are being sampled. Failure to do so can result in wasted time and resources, and ultimately lead to incorrect estimates of the mineral deposit. Therefore, the problem definition stage should be followed by carefully considering the relevant variables to be sampled. Spending a little time and applying some forethought to what needs to be accomplished usually makes the relevance obvious. However, this step should under no circumstances be omitted. Too many studies in the authors' experience have been completed only to realise that the wrong variable had been sampled.

METHODS

Here, the researchers need to consider the framework within which sampling takes place. Some important factors that need to be considered before a method is devised are outlined:

   a. Can we sample the entire population, or do we need to take some subset thereof?
   b. Is the sampling method intrusive and/or destructive in nature?
   c. Is the material to be sampled hazardous in nature?
   d. Is there specialised equipment needed to sample?
   e. Do we use sampling without replacement or with replacement?
   f. What is the required outcome?

Considering all the above, one is now ready to determine the best method for sampling. This is usually accomplished by consulting someone with experience in sampling, specifically of the required kind. This could be as simple a method as pencil and paper, or a shovel and bucket, to sophisticated machinery that requires specialised training. Due care must be exercised about the safety of the sampler at this junction in the project.

TRUTH

In this section, we examine the relative truth of the sampling or subsequent processing of mineral deposits. No process can be considered complete unless the veracity of the sampling methods, collection, and subsequent processing can be determined. This is typically determined via the use of truth tables. In some cases, the results of a truth table require immediate action to rectify a situation; in other cases, the results are '*fait accompli*', and the table is used exclusively to determine the overall truthfulness of the sample.

Consider the example of a mineral deposit survey. To ensure that the sample is relevant, the truth test could be obtained by asking two additional questions:

a. Do the sample and analytical methods used accurately represent the mineral deposit?
b. Are there any factors that could affect the accuracy of the sample or analytical methods used?

Now let us consider the logic behind the truth table we would like to generate. If the answer to both questions is 'yes', we have the following scenario:

i. The sample and analytical methods used accurately represent the mineral deposit; therefore, the sample can be considered truthful.
ii. There are factors that could affect the accuracy of the sample or analytical methods used, but they were properly addressed; therefore, the sample can be considered truthful.
iii. There are factors that could affect the accuracy of the sample or analytical methods used, and they were not properly addressed; therefore, the sample must be considered false.

If the answer to question (a) is 'no', and the answer to question (b) is 'yes', we have the following scenarios:

i. The sample and analytical methods used do not accurately represent the mineral deposit, but factors affecting the accuracy were properly addressed; therefore, the sample can be considered truthful.
ii. The sample and analytical methods used do not accurately represent the mineral deposit, and factors affecting the accuracy were not properly addressed; therefore, the sample must be considered false.

If the answer to question (a) is 'yes', and the answer to question (b) is 'no', we have the following scenario:

i. The sample and analytical methods used accurately represent the mineral deposit, but there are no factors that could affect the accuracy of the sample or analytical methods used; therefore, the sample can be considered truthful.

TABLE 3.1

Truth table for mineral deposit sampling

Scenario	Truth Index for Presence of Mineral Deposit (1 = True, 0 = False)	Truth Index for Size of Mineral Deposit (1 = True, 0 = False)	Truth Probability
1	1	1	0.80
2	1	0	0.10
3	0	1	0.05
4	0	0	0.05

We have a logical contradiction if the answer to question (a) is 'no' and the answer to question (b) is 'no'. In this case, we deal with the sample in the same manner as in scenario (iii). Following is a summary of the results of the truth table (Table 3.1).

In this example, the truth table is used to determine the overall probability of the presence and size of a mineral deposit based on sampling data. The table includes four scenarios based on the results of the sampling:

- Scenario 1: The mineral deposit is present, and its size is accurately determined (truth index = 1 for both variables), with a probability of 0.80.
- Scenario 2: The mineral deposit is present, but its size is inaccurately determined (truth index = 1 for presence, 0 for size), with a probability of 0.10.
- Scenario 3: The mineral deposit is not present, but its size is inaccurately determined (truth index = 0 for presence, 1 for size), with a probability of 0.05.
- Scenario 4: The mineral deposit is not present, and its size is also inaccurately determined (truth index = 0 for both variables), with a probability of 0.05.

The overall probability of the accuracy of the mineral deposit sampling can be determined by summing the probabilities of scenarios 1 and 2, which is $0.80 + 0.10 = 0.90$. The truth table provides a useful tool to evaluate the reliability of mineral deposit sampling and helps to ensure that accurate and relevant data is obtained. Applying the above logic to each mineral deposit sample, one can obtain the overall probability of the truthfulness of the sample. However, truth tables are not limited to mineral deposit sampling. With a bit of forethought and imagination, the truth table can be applied to any form of sampling and is well worth the effort involved in obtaining an indication of the overall veracity of one's results.

CAUSALITY

a. Strict causality

No other factor contributes more to the epistemological basis of sampling than the principle of causality. Here, we are assuming some relationship between the variable we are sampling and the final objective of the sampling process. Under some conditions, this does not require much of a leap of faith, such as the relationship between the grades sampled from an orebody and the resultant overall grade of everything mined within that orebody. Although we cannot say with absolute certainty that this occurs in all circumstances, there is ample empirical evidence that, except for grade variability, which is discussed elsewhere in the document, one can assume that what we sample is representative of what we do not know. Nevertheless, this is not always the case, especially when there is an implicit relationship between the sampled variable and the final objective. Here, consideration must be taken of the dependent and independent variables to ensure that the direction of causality is from the independent to the dependent variable.

Let us consider a case in point. Suppose we want to determine the concentration of carbonate minerals in a copper mineral deposit. The concentration of carbonate minerals can affect the acid consumption during the copper leaching process, which is the final objective of the sampling process. In this case, we can assume a direct relationship between the concentration of carbonate minerals and the acid consumption during leaching. However, we must also ensure that the direction of causality is from the concentration of carbonate minerals to the acid consumption, and not the other way around.

In this case, the direction of causality is obvious; the same cannot be said of many other variables. Consider the case of mining productivity; there is a link between the number of workers and the amount of copper produced. Can one with certainty assume the opposite? That the amount of copper produced increases the number of workers. The assumption is problematic, to say the least. Therefore, it is the project leader's responsibility to steer clear of areas in which the direction of causality cannot be established. In most cases, the direction of causality is not obvious, and caution is needed to avoid erroneous conclusions. The project leader should always ensure that the relationship between the sampled variable and the final objective is established, and that the direction of causality is verified before undertaking any sampling process.

b. Relational causality

In the context of mineral deposits, relational causality refers to the relationship between different variables that may have a modellable relationship rather than a direct cause-and-effect relationship. An example of this is the relationship between the zinc content of different geological units in a deposit and their metallurgical response during processing. Suppose one collects samples from various locations within a zinc deposit and measures the zinc content, as well as the mineralogy, texture, and other characteristics of each sample. One can then use statistical techniques to build a model that relates the zinc content of each geological unit to its metallurgical response during processing, as shown on Table 3.2.

From this, one can see that there is a clear relationship between the zinc content of each geological unit and its metallurgical response during processing. However, the exact nature of this relationship is not immediately

TABLE 3.2

Example of multiple variable relational causalities for zinc geometallurgy

Geological Unit	Zinc Content (%)	Metallurgical Response (% Recovery)
Oxide zone	8.2	73.5
Mixed zone	5.9	62.1
Sulphide zone	3.6	45.6
Deep sulphide zone	1.9	23.9

obvious and may depend on a variety of factors, such as the mineralogy, texture, and other characteristics of each geological unit. By building a model that captures this complex relationship, one can better understand the deposit's metallurgical behaviour and make more informed decisions about how to extract and process the zinc ore.

Nevertheless, we can and do use these types of models effectively in many aspects of our daily lives. How then do we decide what form of relational models we can use? As you may have guessed, the example relationship above has a significant disadvantage. How do we know when the relationship ceases to exist? From this simple example, one can deduce two restraining criteria regarding relational causality.

i.   Certain relationships may have an upper limit to the relationship.
ii. Where the upper limit to a relationship is unknown, only the portion between the lowest and highest samples may be utilised.

THE INFLUENCE OF TIME AND SPACE

One cannot assume that if the sampled relationship today between the ore across a conveyor belt and that retrieved from the plant is at 60%, or the percentage of over-ripe apples delivered to the supermarket is at a similar percentage, that this will be the case in the future and for all plants and supermarkets. Under these conditions, the sample only reflects what happened at a specific location and time. Under certain conditions, where the sampled data exhibits a periodicity, current sampling can be used to predict future events. If the information obtained causes a change in circumstances or action, the periodicity must be ignored, and the variable must be re-sampled. Space also plays a part in determining the relevancy of the sample in another manner; if, for instance, the water consumption in an upmarket area of town is measured per household, one cannot say that this will hold true for rural areas. Due consideration must be taken of the influence of time and space before embarking on a sampling programme if the results obtained are to have any real meaning. Let us look at another example:

The relationship between the grade of nickel concentrate produced from a mineral deposit and the ore grade can vary over time and space, making it necessary to consider the influence of both factors before embarking on a sampling programme. For example, the grade of nickel concentrate produced at a particular location may

TABLE 3.3

Example of the influence of time and space on nickel concentrate grades

Location	Date	Ore Grade (% Ni)	Concentrate Grade (% Ni)
Mine A	1 January 2022	2.3	12.1
Mine A	1 January 2023	2.6	11.8
Mine B	1 January 2022	1.8	10.5
Mine B	1 January 2023	2.1	10.9

be affected by changes in ore quality over time, as well as changes in the processing plant's operating conditions. Similarly, the grade of nickel concentrate produced at one mine may differ from that produced at another mine, even if the two mines are located in the same region as shown in Table 3.3.

From Table 3.3, the relationship between ore grade and concentrate grade varies both over time and across different locations. At Mine A, the ore grade increased from 2.3% to 2.6% Ni between 1 January 2022 and 1 January 2023, while the concentrate grade decreased slightly from 12.1% to 11.8% Ni. At Mine B, the ore grade increased from 1.8% to 2.1% Ni over the same period, and the concentrate grade increased from 10.5% to 10.9% Ni. These variations illustrate the importance of considering the influence of time and space when designing a sampling programme for a mineral deposit.

REDUCTION OF BIAS

According to the Oxford dictionary, bias is a predisposition or inkling towards prejudicial or ignoring of a factor that needs to be considered. Of all the human traits, this one encompasses the human spirit more than any other and has done more to the detriment of science and progress than any other factor. For these reasons, reducing bias must hold pride of place within the context of correct sampling procedures. Too often, pet ideas and projects gain the upper hand over rational thinking and empirical evidence, resulting in biased sampling. One would normally not expect the subject of ethics to be contained within the context of sampling practice.

Nevertheless, this is an extremely important part of the sampling process, and practices that limit the amount of personal bias inherent within the sampling process need to be emplaced. Suppose a pharmaceutical company had produced a particular drug, and the person responsible for the sampling of results decided that as the tremendous market potential was the Asian market, s/he was going to limit the test subjects to exclusively Asians. Although this financial logic cannot be flawed, because this is the largest market, the potential financial loss from being sued due to the drug's unfavourable effects on other population groups is being ignored. A disproportionate amount of research has been conducted in this area on limiting bias inherent in sampling methods. These methods attack either the problem of personal bias via methods of ensuring randomness of the sample or reducing bias introduced by the sampling method itself. Yes, this does occur and is perfectly acceptable if the sampler is aware of the limitations and considers them or compensates for them.

REPRESENTIVITY

This is the most contentious of all the sampling factors. Moreover, this also results in the area in which the most significant potential for problems in sampling methods is evident. How do we define representivity? The easy answer is to obtain a sample of all the relevant parameters, a sample in sufficient number, and a wide enough spread to ensure that the sample is representative of the population that will be inferred. One is immediately struck by some questions that arise from the above statement:

    i. How do you know what all the parameters are before sampling?
    ii. What are sufficient samples?
    iii. How do you determine if they are using a wide enough spread?

We do not know the answers to these questions. However, all is not lost; with some reasoning, one can infer the parameters to a greater extent. One may have a project already completed of a similar nature. From these results and subsequent follow-up, one may infer certain *a priori* information, which may benefit the project. Additionally, you need to apply a judicious dose of reason. Consider a survey in which a certain maize (or corn) meal producer wishes to determine the potential for a new maize meal. It will undoubtedly help if people are stationed at local supermarkets with samples of cooked maize meals. Here, the person conducting the survey can ask whether the taster enjoys the sample. This will determine acceptability. This, however, does not determine whether the individual will purchase the product. This question will have to be asked explicitly. At this stage, we still do not have sufficient information to determine whether the product will be a success, as the extent of consumption has not yet been identified. Once again, we will have to ask explicitly how often maize meal is consumed within the household and how many members exist within the household. You can now see that the question of parameters to sample has been dealt with using a reasoned approach and now represents the individual's potential.

The answer as to how many samples you take is a simple one. Take as many samples as required until your analysis results do not change substantially. This is a subjective choice and is determined by the risk one is prepared to take. Suppose the average result changes by only 5% by adding additional samples, and you or the customer are prepared to accept a 5% risk. In that case, you can stop sampling as one has sufficient samples.

When considering whether one is using a wide enough spread, reason must prevail. Let us consider the previous example once again. Suppose we took our samples in an upmarket area such as Houghton in central Johannesburg (South Africa); our survey results would most probably show that there is not much of a market for the new maize meal. Alternatively, if the survey were conducted exclusively in Soweto (Township, southwest of Johannesburg), you could easily conclude that the new product would succeed tremendously. This is because one does not consider the varying demographics when extrapolating the sample to the population. Ideally, you need to cover all demographics, populations, and financial groups to achieve representivity. Furthermore, you need to remember that once all groups have been sampled, the overall result still needs to be weighted by the population within each group.

TABLE 3.4

Example binary classification of statistical results

Location	Condition	Condition
Test	True	False
Condition	True positive	False negative
	False positive	True negative

ACCURACY, PRECISION, AND MEASUREMENT RESOLUTION

Accuracy, precision, and measurement resolution are critical considerations in any sampling procedure. Accuracy refers to how close the estimated value is to the actual value, while precision measures the reproducibility of the estimate. If the estimates are consistently biased, the results may be precise but not accurate. It is important to note that accuracy should not be confused with measurement resolution, which is the smallest quantity that can produce a change in the estimate. Additionally, precision should be related to the level of measurement, as extraneous precision can be useless.

Binary classification is a common method for assessing precision and accuracy. Table 3.4 illustrates a hypothetical example of this classification, using the regression function of initial estimates for a gold deposit. Estimates that require an increase are labelled as positives, while those that require a decrease are labelled as negatives.

The accuracy of the regression can be determined using the formula:

$$\text{Accuracy} = \frac{\text{Number of True Positives} + \text{Number of True Negatives}}{\text{Total Number of Estimates}} \tag{3.1}$$

Precision is obtained by:

$$\text{Precision} = \frac{\text{Number of True Positives}}{\text{Total Number of True Positives and False Positives}} \tag{3.2}$$

TYPES OF SAMPLING

In the most straightforward case, for example, taking a whole batch of production from an assembly line, it is possible to sample every item in the population. However, in most cases, this is not possible. There is no way to sample all oranges in the set of all oranges. Sampling methods are essentially divided into two categories with several subcategories.

Probability sampling:

a. Systematic sampling – This method involves selecting a fixed interval of ore samples from a defined population of ore blocks. For example, every fifth or tenth block of ore may be sampled, ensuring that a representative sample is obtained across the entire population of blocks. Systemic sampling selects samples at a predefined interval, say 5 m. It is particularly effective in cases where human intervention in the order of the occurrence of samples is not an issue. On the opposite side, if there is a periodicity within the population and it is some factor of the sampling interval, you could end up with a biased sample.

b. Simple random sampling – This involves selecting ore samples randomly from the population of ore blocks. Each block has an equal chance of being selected, ensuring a representative sample.

c. Cluster sampling – In this method, the ore blocks are divided into groups or clusters, and then a random sample of clusters is selected for testing. For example, a mine may be divided into different zones, and then ore blocks from each zone are selected randomly for testing. A certain area is chosen with cluster sampling, and a group of samples is taken within this area. This can dramatically reduce the cost of the exercise; however, the methodology needs to be copiously documented to ensure that the sample does not get taken in the incorrect context sometime in the future.

d. Stratified sampling – This involves dividing the ore blocks into strata or layers based on certain characteristics (e.g. ore grade or mineralogy) and then taking random samples from each stratum. This ensures that the sample is representative of the entire ore body. Stratified sampling is essentially a process of domaining and then sampling each homogenous domain within its right. Care must be taken when combining domains if this methodology is used to ensure each domain has a representative weight. This methodology is used to ensure homogeneity and reduce variability.

Non-probability sampling:

a. Quota sampling – In this method, a certain number of ore samples are taken from each stratum or zone, regardless of their representativeness. For example, a certain number of samples may be taken from each level of the mine, regardless of the ore quality. The target group may be subdivided in quota sampling into perceived homogenous subgroups. The sampling leader then decides on a quota of samples required to form each group based on his/her judgement. However, in a quota system, the decision of whether to sample or not is non-random and can result in bias. In the authors' opinion, this is one of the least effective methods.

b. Accidental sampling/grab or convenience sampling – This method involves taking ore samples from whichever blocks or areas are easily accessible, without regard to their representivity. This is generally not a recommended method as it may lead to biased results, but it can be useful in situations where time or resources are limited. This is the least effective sampling method, as there are no controls on opinion or how the sample is obtained. At best, this should be used to determine whether additional work needs to be done within a certain area.

ACTUAL SAMPLING PRACTICES ON A MINE

Now that you have an idea of what types of sampling there are, and the rules associated with such sampling let us now consider how sampling needs to take place on a mine.

   a. A sampling of the exposed area of an orebody

      Here, we are considering an area where the orebody is directly exposed to the sampler. These can be outcrops, exposed underground mining faces, benches in an opencast, gravel deposits, etc. Each of these will be considered. However, except for gravel beds, reverse circulation, rotary air drilling and surface rock dump, the overall sampling method of exposed rock *in situ* remains the same. We will initially deal with the sampling of gravel beds (because of some similarities), reverse circulation and rotary air drilling, followed by rock dumps and then the *in situ* sampling method.

   b. A sampling of gravel beds

      Here, it is simple to accomplish a non–biased sampling procedure that is representative of the area of interest, if not the whole. This is accomplished by utilising an equidistant sampling interval and sampling normal to the banks and across the entire bed. Having said this, other factors beleaguer an effective sampling practice of this kind.

- Unequal proportions when crossing the gravel bed: Here, we are not referring to the differential depth that occurs naturally and hence is representative of the volume. Instead, the unequal width of the sample, which can occur as a result of using a back actor and not controlling the sides of the excavation, causes oversampling at that position. This could have particularly deleterious effects, especially if the gravel beds had an original meandering nature. This could produce concentrations of mineral content along bends, which, if oversampled, will lead to an over-evaluation of that portion of the deposit. For this reason, the preferred method of sampling is to embed two planks or steel plates of the correct width apart before sampling and then remove the material between the restraining structure.

- Transport of sample to the area where sample reduction takes place: Here, multiple errors may occur, listed as follows:
  - i. Transport container not properly emptied and cleaned between trips.
  - ii. Loss of sample during transportation.

The sample preparation area was not adequately cleaned between channel samplings.

- Sample volume reduction: Once again, you may encounter multiple sources of error. Due to the large sample size produced under most circumstances when doing this type of sample, an efficient and unbiased reduction method is required. Here the time-honoured method of coning and quartering, if

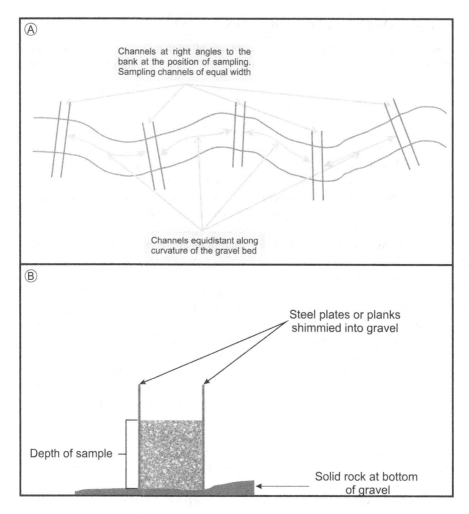

FIGURE 3.1   (a) Example of sampling layout for gravel beds, and (b) section of the sample layout.

done correctly, is an efficient methodology. Following are the points to watch out for:

i.   Not sufficiently mixing the sample during coning: It is important to thoroughly mix the sample while coning, especially in the initial cone.

ii. Allowing material from outside the sample to be included in the sample: This is why, if possible, a clean concrete pad can be utilised for coning and quartering. If this is not possible, a thick plastic sheet can be placed on hard ground that is as flat as possible; however, due care must be exercised so that during the process, the sheet is not damaged and material from below the sheet is included in the sample.

iii. Lack of supervision in which a portion is taken before proper coning and quartering is accomplished.

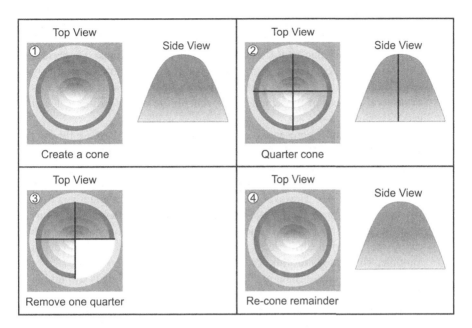

FIGURE 3.2 Illustration showing the method for coning and quartering to reduce gross sample for subsampling. First, shuffle the sample at least three times: (a) the gross sample is first piled into a cone and flattened looking down from above, (b) divide the flattened cone into half, (c) the flattened cone is divided into four quarters, (e) opposite quarters taken for mining and forming, and (d) re-cone to repeat the process.

To summarise this method, refer to Figure 3.1.

- Select equal distance intervals along gravel beds.
- Choose a standard sample width and stick to this throughout the sampling.
- Sample across the entire bed, at right angles to the bank.
- Transport to the area where coning and quartering are to take place.
- Once the entire section has been sampled:
  i.   Thoroughly mix the sample and shape it into a cone (Figure 3.2).
  ii. Using a plank or shovel, split the cone into four equal portions.
  iii. Take one portion away.
  iv. Thoroughly remix the remaining three portions.
  v.   Redo points i and ii until the sample has been reduced to a manageable size.
  vi. Collect the entire sample, including fines.
  vii. Bag, label and tightly close to prevent contamination.

REVERSE CIRCULATION (RC) DRILLING

Although this type of drilling is much cleaner than other types, and more of the sample is retained, there are a few matters that you need to consider when using this method to obtain a sample.

- Friability of ground being drilled: If the ground being drilled is highly friable, the ratio of dust to rock coarse particles may be high, which can lead to mass loss due to the escape of dust. With less mass for certain mineral content, over-evaluation of assay results can occur. If a relatively large proportion of the ground drilled is of a liberation size or less, loss of mineral content of the sample can occur in the dust, leading to lower grades assayed.
- Obtaining a sample representative of the distance drilled: If you only want an average value of the ore body across the entire distance drilled, it is a simple matter to retain the sample/s in the original bag and, if necessary, in a clean environment, do a cone and quarter exercise to reduce the sample to a manageable size. However, if the samples must be representative of the whole drilling composite length, you face an entirely different set of circumstances. To satisfy this criterion, it is not sufficient to change the sample capture bag when the driller calls out that they have reached a certain depth multiple. The reasons for this are:
  i.   Sample loss due to changing the sample bag while drilling is still taking place:
  ii. Because of the inhomogeneous sample size obtained during drilling and thus progressing up the inner tube, smaller particle size and dust will travel faster than coarse rock particles, making it entirely possible, if not probable, that sample from the previous depth section will still be in the inner tube while drilling the next depth section. This will lead to the incorrect value being attributed to each and every depth section. For this reason, drilling needs to be stopped if a sample being a specific depth interval is required. All material in the inner tube should be expunged and collected in a separate sample bag before drilling is allowed to recommence.

ROTARY AIR DRILLING

A similar approach must be used here as with RC drilling; however, because this method generates a lot more dust, extra care must be taken to ensure that all minute dust particles are collected between sample depth intervals (Kahraman, 1999). Before beginning drilling, as with RC drilling, the hole must be expunged, all samples collected, and the area must be cleaned up. As a precaution, under no circumstances should RC or rotary air drilling be seen as accurate as core drilling or direct sample selection, as in the case of direct channel chip sampling.

CORE DRILLING

Here one is obtaining a 'hopefully' solid sample of the drilled area, which should provide a far more accurately representative sample if a few simple rules are followed. These rules are a function of the information required to provide a correct transformation of the drilling as obtained to the ideal representation of such drilling. If the core-bedding angle is not taken, one requires the dip and direction of the borehole and the direction of the strike (Kumar et al., 2019). If the core-bedding angle has been obtained, the dip and direction of the borehole and the direction of strike can be used as a means of checking. When obtaining the core-bedding angle, it is of utmost importance that a consistent singular method is utilised as there are two methods of measuring this angle which provides complementary angles and hence will not produce the same results.

THE TWO METHODS OF MEASURING CORE-BEDDING ANGLE

There are two methods that can be used to measure the core-bedding angle. The first method requires that the core-bedding angle be measured along the direction of drilling while the second method involves measuring the core-bedding angle at 90° to the direction of drilling, as illustrated in Figure 3.3.

To survey or not to survey is governed by multiple factors such as the depth of the hole, the hardness of the rock, the homogeneity of the rock, stratification, and the inclination at which the hole transgresses the strata and finally, the speed of drilling. However, having said this, unless the rock is very soft and one is drilling at 90° to the strata, it is undesirable to drill more than 50 m without utilising a subsequent survey of the hole. Correcting widths to be normal to the orebody is of extreme importance. At times, core-bedding angles may confuse matters, mainly if the angle on the upper and lower contact differ significantly. Here it is extremely valuable if the true thickness can be calculated via an alternative means, which would not be possible if the hole had not been surveyed.

As previously stated, if the collar and the depth of the intersection hole are not carefully obtained, the position of the hole will be incorrect, and thus the information required to obtain the correct width may also be suspect. In this case, should we drill a second hole? Although a second and third deflection off the original hole is to be commended, if the hole is extremely long, one or two long deflections are also recommended. You must get one's money's worth, and this is the best way to achieve it. Firstly, the short deflections indicate short-range variability, and the long deflection provides additional sample value at further distances. This can be highly beneficial if you are attempting to obtain a higher level of resolution and idea of variability further away from dense data. Drilling a second hole to firm up the grade is an entirely different kettle of fish. Here, you intentionally invoke bias in the sample dataset by not adhering to the underlying random sampling prerequisite. Furthermore, unless there is a shallow nugget effect or the orebody variation is unduly low, the second hole to prove the grade will do nothing of the sort at an unneeded expense to the company. Finally, the resultant cluster will provide a problem for the modeller to fix without the benefit of providing additional beneficial information.

We now move on to selecting samples and documenting the core once the hole has been drilled. Here it is the authors' opinion that the visit to the core yard for logging and sampling of the core be accomplished by at least two geologists. It is judicious to remember that the core is only a few millimetres in diameter, typically from 32 mm to 60 mm. Thus, one does not have the advantage of seeing on either side of the area of the sample as one might when observing a face or an outcrop. Hence, you may log

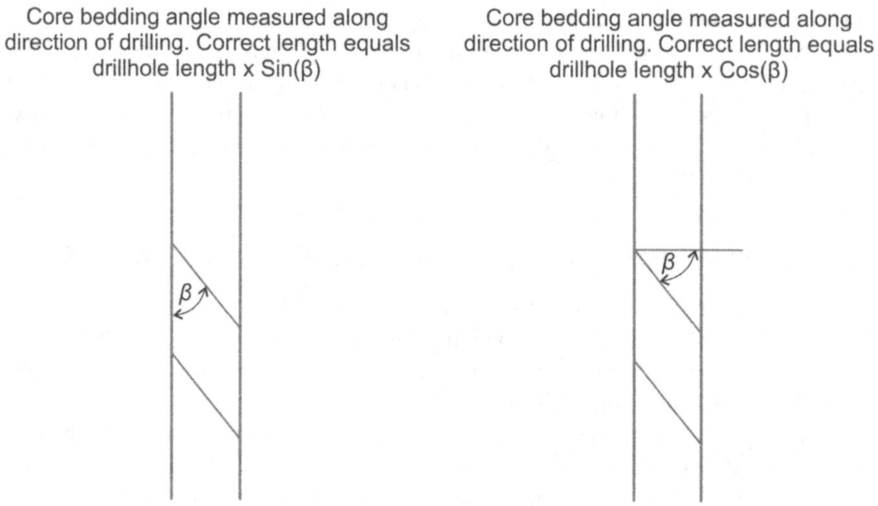

Core bedding angle measured along direction of drilling. Correct length equals drillhole length x Sin(β)

Core bedding angle measured along direction of drilling. Correct length equals drillhole length x Cos(β)

FIGURE 3.3   Example of two methods for measuring core-bedding angle.

a portion of a hole based on the expectation of a certain rock type, whereas if there had been a second opinion, different logging would occur. Too often, the authors have observed a peculiarity in the data that was unexplainable at first observation but later solved when going back and re-observing the core. A few factors control the question as to whether to cut the core in half. These are:

- Diameter of the core.
- Sample width.
- Size of aliquot used by the lab.

One would like to have enough samples for three submissions, the initial sample, a check assay if necessary, and an assay from an additional laboratory if necessary. Suppose that one has a core diameter of 30 mm and a sample width of 10 cm, and the *in situ* relative density is of the order of 2.7 t/m^3. If you were to halve the core, the total mass of the sample would be of the order of 70 g. Now consider if the aliquot used by the laboratory is 50 g; if required, additional checks would not be possible. Now consider if the sample width were 25 cm, you would obtain approximately 176 g of sample, undoubtedly sufficient for a sample to check and an alternative laboratory submission. It is not difficult to imagine a situation where you would have to utilise the second half of the core if queries as to the veracity of the assay were to arise. The problem under these circumstances is that the second half of the core is no longer available for re-inspection, given that very often, it is not known *a priori* as to whether the second half of the core will be needed for re-assay. It is, therefore, a diplomatic protocol to ensure that high-quality photographs be taken of the core and stored securely in a database that regularly gets backed up; additionally, these photographs need to be indexed in such a fashion that retrieval thereof is almost immediate.

CHIP CHANNEL SAMPLING

a. Equipment required:
- Hammers
- Chisels
- Tapes (5 m, 30 m, and 60 m). Used for locating the position of samples
- Marking crayons
- Sample bags
- Tickets with a unique numbering and no duplicate numbers
- Sampling dishes must be clean and orderly with no holes or cracks
- Carrying bags
- One clino rule
- Wire or nylon brushes
- A bucket with clean water
- Field book and pencil
- Spray Paint
- Goggles
- Gloves
- First aid pouch

b. Sampling Procedure

Locating sample sections

  i. Stoping

Position your first section, half the sampling interval, from either the top corner of the face or at the start of the sampling area.

Fix the position of the first and last sections by tape triangulation or distance and direction from a fixed point, for example, a numbered survey peg. The positions of the first and the last section must be physically plotted on the plan and in the section sheet to check that the correct peg, dip, and bearing were used and that the correct position has been obtained.

At least one section should be fully sampled. The reef and the hanging wall, footwall, and internal quartzite should be sampled. All attempts should be made to ensure that the reef samples are segregated, considering apparent differences in quality, pebble size, and supporting matrix.

 ii. Development:

Locate the last section of the previous sample from a fixed position and then progress the standard sampling interval to position the first sample.

Mark off sample section positions to be taken at equal standard sampling distances.

Locate the last section from a fixed point, such as a survey peg.

The positions of the first and the last section must be physically plotted on the plan and in the section sheet to check that the correct peg, dip, and bearing were used and that the correct position has been obtained.

At least one section done on the day should be a fully sampled section, which is the reef; hanging wall, footwall, and internal quartzite are also being sampled. All attempts should be made to ensure that the reef samples are segregated, considering apparent differences in quality, pebble size, and supporting matrix.

Dress down the section and ensure loose pieces of rock on the face are removed, then wash the sidewall from top to bottom at every sampling section with clean water.

i. If underground on a slope, make it safe and test the hanging wall and face; if in any doubt, call the miner in charge to make it safe. Additionally, check for misfires and sockets. If necessary, move the section to a safe area and adjust your sampling section positions accordingly. Finally, brush the face with a wire or nylon brush to remove sticking fines or oxidisation, which may have occurred in old workings.

c. Marking where the samples are to be taken:
- Draw a section line from the hanging wall, at right angles to dip, to the footwall at the position where the sample is to be taken. Now draw two control lines parallel (10 cm apart is a good standard; however, wider samples may be taken) on either side of the section line. Identify the reef contacts on the hanging wall and footwall. Now draw two lines, the first 1 cm above the hanging wall contact and the second 1 cm below the footwall contact, parallel to the hanging wall and footwall contacts.
- Classify the rest of the reef according to appearance and draw lines parallel to the contact.
- Mark off samples from the bottom upwards.
- Try and keep the maximum sample width less than 25 cm.
- Where high values are expected (top or bottom loading), the sample should be segregated accordingly.
- Check samples to be taken wherever the reef appears to have higher values (e.g. abundant pyrite mineralisation in gold deposits). These are taken directly adjacent to the original sample and of the same width as the original. It is highly recommended that a full check section be done on each day's work. This has the benefit of providing the most accurate nugget effect. Assuming, of course, that the information is stored in a manner that can be retrieved and utilised.
- For Waste on Contact, the sample should be 2 cm above and 2 cm below the contact.
- All precautions must be taken to ensure a mass of greater than 0.3 kg and sufficient sample for a re-assay and check assay if required.
- Sample all exposed reef bands.
- Ensure that any sample widths do not cross reef classifications/mineralisation.
- After the sample area has been marked off, it is strongly recommended to spray the area with the paint and again connect the demarcation lines. Upon finishing sampling, no paint should show between the demarcated sampling area.

d. Measuring sample width
  - Determine the true dip and strike for every section.
  - Measure widths correctly and at right angles to the plane of the reef, that is, parallel to the guidelines on strike; book widths correct to the nearest 1 cm, in sequence, from the bottom upwards. Now check the measurement of channel width. This must equal the sum of individual widths.

e. Allocation of tickets

  Start assigning the sample ticket numbers from the bottom upwards, ensuring that the numbers allocated are recorded correctly.

f. Measuring a stoping width

  Measure the stoping width at each section, taking the width between 1.5 m and not more than 3.0 m from the face. This ensures that the full width of the excavation is taken after potential subsequent scaling and not merely the face width. Measure stoping width at right angles to true dip and strike, the shortest distance between hanging wall and footwall.

g. Obtaining the true dip and strike
  i. Dip

    Fix one end of a tape, approximately 2 m long, on a hanging wall and move the other end of the tape on the same plane until the position is found where the maximum dip is obtained. Now, measure and record the true dip of the reef.

  ii. Strike

    Fix the tap at one end and stretch the tape at the other end; once the tape is level, the strike direction has been obtained on the plane of the reef. The true dip direction is at right angles to the strike direction.

h. Chipping, numbering, and collecting samples

  Remember, accidents can happen at any time so maintain safety awareness constantly.

  i. Use clean tools and equipment.
    - Chip sample in the correct sequence – from bottom upward. This ensures that no contamination occurs of samples below these being chipped from the fines above.
    - Hold the sample dish firmly against the face directly below the chipped sample.
    - Shield the chisel with a clean glove or canvas bag.
    - Hold the chisel in hand with the palm pointing upwards.
    - Chip footwalls contact sample first.
    - Chip to an even depth throughout the sample area.
    - Transfer sample and corresponding ticket from dish to bag.
    - Clean the dish after every sample with clean water.
    - Discard and re-chip any spilt or contaminated sample.
    - The sample bag must be fastened securely and placed in a secure transport bag.
    - Mapping of Reef between sections.
    - Correlate the reef between sections, drawing a representative sketch of the face.

- Measure and record any reef in the hanging wall or footwall.
- Measure and record off-reef portions.
- Measure and record all geological features, that is, dyke/fault positions, dips, throws, and strikes, including all sedimentological parameters.

j. Underground workings

Here it is of benefit that, in addition to sampling, the sampler

- Examines sweepings behind the blasting barricade.
- Records and measures excessive accumulations and takes samples per panel.
- Examines packs for fines.
- Checks the condition of the blasting barricade and/or scatter pile.
- Checks for reefs in the hanging wall or footwall and report between samplings.
- Reports if blasting barricades are not being used.
- Finishing.
- Check that all equipment is available and stored in bags.
- Check the total number of samples against the number in the field book.

k. Field book entries

- It must be neat, with errors crossed out with a single line.
- Sketches and diagrams should be such that any other person can capture the data if necessary.

l. Underground samples were taken, and the conveyance thereof from the workplace to the Assay Laboratory:

- Chip mine samples as per the required sampling standard.
- Bag and label samples as per the required sampling standard.
- Count the number of samples chipped at the bottom of the stope and compare it to the number of samples recorded in the sampler's field book.
- Ensure all samples are securely bagged in a carry bag to be conveyed to the surface sampling store.
- Samples chipped underground must be counted again on surface in the sample store, and the totals received should be compared to ensure no samples are lost or unaccounted for.
- A waybill book should be filled in and reflect the correct number of samples.
- The waybill must accompany the specific batch of samples to the Assay laboratories.
- Samples must be securely locked in a sealed container when transiting to the Assay Laboratory. This is done to ensure that the integrity of the samples transported is maintained and no tampering occurs, such as salting of samples (and yes, this does occur).
- At the Assay Laboratory, samples are counted again and compared to quantities recorded per waybill book.
- Samples are received and signed off by Assay Laboratory personnel.
- Any missing or unaccounted sample must be investigated and reported without delay.

m. Belt samples were taken, and the conveyance thereof from the site to Assay Laboratory:
  • Utilise correct tonnage sampling interval.
  • When using a Go Belt sampler, the rubber sample sweeper on the Go Belt sampler should be checked at least once a week and adjusted as needed.
  • Selective sampling of the belt must be avoided at all costs, as this introduces value bias.
  • The weight metre must be calibrated at a minimum interval of once a month.
  • Control of belt and weight metre to be handled only by an authorised technician.
  • Samples are to be collected in a metallic bin.
  • Bin to be locked before dispatching from the site.
  • The driver must be in possession of a waybill.
  • Upon arrival at the Assay Laboratory, samples are received and signed off by Assay Laboratory personnel.
  • Mass of sample to be measured and documented upon arrival at the Assay Laboratory.
  • Mass of final sample selection and the number of samples selected for assay to be scrupulously documented.

THEORY OF SAMPLING

The theory of sampling (TOS) is a crucial framework aimed at reducing errors in the grade control process. It addresses two fundamental questions: how to select a sample and how much material to take. TOS acknowledges various sampling errors that can occur from sample collection to assaying, introducing uncertainty in the final assay value. The pioneer of TOS was Pierre Maurice Gy, a French Mineral Processing Engineer, who recognised the need for precise sampling methods in the late 1940s when faced with sampling a massive heap of lead concentrate with blocks ranging from several tonnes to microgram particles. The foundational TOS was developed in the early 1950s through to the 1970s.

An important aspect that TOS deals with is the 'nugget effect', which quantifies the natural variability between samples taken at small separation distances. Geological and sampling factors influence this effect and arise due to inadequate sample sizes, suboptimal sample collection, and preparation methods. To improve data accuracy and reduce extreme data values, sampling protocols throughout the mining process should be optimised to minimise the nugget effect and overall data skewness.

To understand the overall sampling error, TOS breaks it down into several components identified by Gy (1982) and Pitard (2006):

$$TSE = \{INE + FSE + GSE + PIE1 + PIE2 + PIE3\} + \{IDE + IEE + IWE + IPE + AE\}$$

TABLE 3.5

Summary of origins and nature of errors

Origin of Errors	Nature of Errors	Identity of Error	Cause of Error	Range of Error (%)
Particulate nature of ore	Distribution of mineral in host rock	*In Situ* Nugget Effect (INE)	Clustering of small gold grains/ occurrence of large grains	50–100
	Compositional heterogeneity	Fundamental Sampling Error (FSE)	Particulate nature of the ore	
	Distributional heterogeneity	Grouping and Segregation Error (GSE)	Inhomogeneous mixing	
Sampling- and subsampling equipment	Geometry of outlined increment is not recovered	Increment Delimitation Error (IDE)	Incorrect cutter design	10–20
	Portion extracted is not the same as delimited increment	Increment Extraction Error (IEE)	Incorrect cutter speed	
Handling of samples and subsamples	Non-random variation after extraction	Increment Preparation Error (IPE)	Sum of variances introduced by handling	
	Proportional sampling	Increment Weighting Error (IWE)	Uneven flow of ore	
Type of sampling process	Small scale variability	Process Integration Error (PIE1)	Variability due to sampling method used	0.1–4
	Large scale non-periodic sampling variability	Process Integration Error (PIE2)	Variability due to process cycles	
	Large scale periodic sampling variability	Process Integration Error (PIE3)	Variability due to periodic heterogeneity	
Laboratory	Analytical technique	Analytical Error (AE)	Lack of diligence in procedural steps/ incorrect procedure	

See Table 3.5 for the explanation of each abbreviated identity of error. The first six components are random errors, which cannot be entirely eliminated but can be minimised. The last five components represent sources of bias, which can be eliminated with appropriate measures.

Fundamental Sampling Error (FSE) is a critical error in the TOS that relates to compositional heterogeneity, such as variations in grade, within a material. It plays a crucial role in grade control as it can lead to the misclassification of ore and waste. To manage FSE and reduce uncertainty in material classification, optimising sample mass and size reduction processes is essential. Minnitt et al. (2007a) broke down FSE into its components, with FSE variance being the only error that can be estimated before conducting the actual sampling. FSE represents the smallest achievable residual average error inherent in the sample due to the physical and chemical composition and particle size distribution (François-Bongarçon, 2002). It arises from two characteristics of broken ore materials:

a. Compositional heterogeneity refers to differences in the internal composition between individual fragments of sampled material, resulting from how they are structured.
b. Distributional heterogeneity: This represents variations in the average composition of the lot from one position to the next in the lot. It is responsible for the irregular distribution of grades in groups of fragments of broken ore.

Important points about FSE include:

a. It cannot be entirely eliminated because most ores are not uniformly structured or composed throughout; they are heterogeneous, even at the molecular level (François-Bongarçon, 2002).
b. Minnitt et al. (2007a) focused on three fundamental problems related to sampling concerning FSE:
    i.   Was the error introduced when a sample of a given weight, M_s, was taken from a pile of broken ore?
    ii. Determining the sample weight is required to ensure that the sampling error does not exceed a specified variance.
    iii. Identifying the degree of crushing and splitting needed to achieve a specified value for the error variance.

Components: Gy (1982) provides an equation to express the variance of the fundamental error when the mass of a pile of broken ore (M_l) (i) is significantly larger than the sample mass (M_s). The equation is given as:

$$\sigma_{FSE}^2 = \frac{f \times g \times c \times l \times d_N^3}{M_s}$$

- f = Shape factor – relates volume and particle diameter to one another = 0.5
- g = Granulometric factor/fragment size distribution factor = 0.25
- c = Mineralogical composition factor – a product of density and large relative variance = 16,000,000
- l = Liberation factor – dimensionless number – ranging between 0 (not liberated) to 1 (completely liberated)
- d_N^3 = Nominal size of fragments
- M_s = Mass of sample (g)

Simplifying Gy's (1982) equation, when the complex written as *fgcl* is made equal to *K*, the equation becomes:

$$\sigma_{\text{FSE}}^2 = \frac{k \times d_N^3}{M_s}$$

When the equation *fgcl* is made equal to a constant value *K*, we can answer the following questions:

1. Error when a sample of given mass is taken: The error introduced when a sample of given mass is taken can be expressed by the equation *fgcl* = *K*, or in a simplified form:

$$\sigma_{\text{FSE}}^2 = \frac{k \times d_N^3}{M_s}$$

2. Mass of sample required – sampling error not to exceed specified variance: The mass of the sample required to ensure that the sampling error does not exceed a specified variance can be determined using the equation *fgcl* = *K*, or:

$$M_s = \frac{k \times d_N^3}{\sigma_{\text{FSE}}^2}$$

3. Crushing and splitting to achieve a specified value for the error variance: The equation *fgcl* = *K* can be used to find the crushing and splitting requirements needed to achieve a specified value for the error variance, as shown by:

$$d_N^3 = \frac{M_s \times \sigma_{\text{FSE}}^2}{k}$$

Equation enhanced: Gy initially used the cube (3) of the nominal top size d_N in the equation, but further work suggested that its value could be a function of the ore type rather than a constant value of three. François-Bongarçon (1993, 1995) provided a new understanding and approach to the interpretation of paired samples. It was established that the exponent (alpha) and the value for *K* could be calibrated, allowing the calculation of FSE for any given mass at any given nominal fragment size. By using *K* and alpha, the liberation size could be determined, and the FSE was found to be a function of the grade of the ores sampled.

Nomogram: Minnitt et al. (2007b) described an experiment and how to determine a sampling protocol that ensures the FSE does not exceed a predetermined precision at any stage in the sampling procedure. A linear graph is used to calibrate alpha and *K*, aiding in the determination of an appropriate sampling approach, as shown in Figures 3.4 and 3.5.

Three modes of sampling

François-Bongarçon (2002) identifies three sampling modes corresponding to different types of automatic samplers used in a process plant where one-dimensional lots,

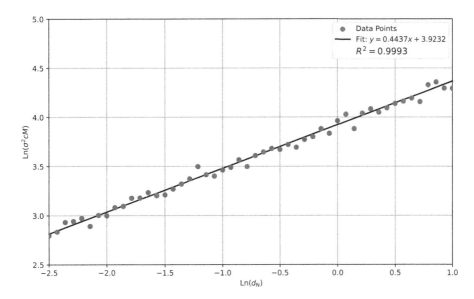

FIGURE 3.4   A linear graph to calibrate alpha and K to determine sampling approach.

such as broken ore on conveyors and slurry streams, are sampled. These modes are as follows:

a. Taking part of the flow part of the time: This involves samplers like internal pipe bleeders, injector- or poppet samplers, which extract a portion of the flow intermittently.

b. Taking part of the flow all of the time: This mode includes samplers like in-pipe derivations, pressure bleeders, or chute discharge derivations, which continuously extract a portion of the flow.

c. Taking all of the flow part of the time: This mode utilises samplers like go-belt- or cross-stream samplers, which intermittently collect the entire flow.

Among the three modes of sampling, the correct approach is to take all the flow part of the time, using samplers like go-belt- or cross-stream samplers.

Hidden costs of sampling errors

Minnitt (2007) discusses the significant impact that sampling errors can have on costs and decision-making. Some examples include:

a. A bad protocol incorrectly implemented for blast hole sampling cost a mine US$134 million over a ten-year period (Carrasco et al., 2004).

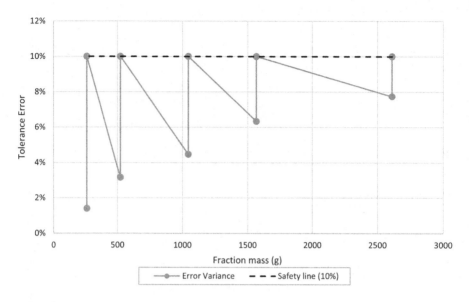

FIGURE 3.5   A plot of the FSE of each step.

b. Incorrect sampling of tailings in a flotation plant resulted in a cost of US$2 billion over a 20-year period (Carrasco, 2003).

c. Another case study by Carrasco et al. (2004), mentioned by Minnitt (2007), illustrates the benefits of installing a sampling station for tailings grade and pulp density in a large copper operation in Chile. The installation revealed that the traditional metallurgical balance of 0.15% Cu was too low by 0.05% Cu. The daily tailings discharge of 96,000 tonnes resulted in the unaccounted loss of approximately 17,500 tonnes of copper every year over an 87-year period, leading to a US$2.2 billion loss for the mine.

THE IMPORTANCE OF TOS AND PROPER SAMPLING PROTOCOLS

The TOS should always drive sampling decisions and the establishment of sampling protocols. It is essential to understand, minimise, and eliminate errors introduced during sample collection and analysis. Only samplers that comply with the third mode of sampling, taking all of the flow part of the time, are correct (Pitard, 2005). A sample is considered correct if all the fragments in the bulk to be sampled have an equal probability of being selected in the sample (Gy, 1982). Using 'cheap' and structurally biased samplers can lead to the production of biased samples and significant unaccounted losses with substantial financial implications. To ensure reliable data and accurate decision-making, the implementation of proper and accurate sampling methods is crucial.

CONCLUSION

The accuracy and reliability of resource estimates heavily rely on the appropriate selection and application of sampling methods. The chapter explored the principles underlying the sampling process in the mineral resources industry and examined the various types of sampling methods that are widely used. The significance of representative sampling is at centre of any sampling method. It is important to ensure that collected samples accurately represent the mineral deposit, in terms of its variability and intrinsic characteristics. To design effective sampling strategy, understanding mineralisation's geological context and spatial distribution is crucial. Conventional and advanced techniques of sampling were presented in this chapter. Examples of conventional methods include grab sampling, channel sampling, and trench sampling, and they were extensively discussed, examining their strengths, limitations, and suitable applications. Further, the chapter explored more sophisticated techniques, including systematic sampling, which aim to enhance sampling efficiency and data reliability. Sampling is a fundamental component of mineral resource estimation, as it provides essential data for understanding the characteristics and quality of a mineral deposit. Therefore, the comprehensive examination of sampling principles and techniques provided invaluable insights for professionals involved in mineral resource estimation, including geologists, mining engineers, and resource geologists.

REFERENCES

Carrasco, P., Carrasco, P., & Jara, E. (2004). The economic impact of correct sampling and analysis practices in the copper mining industry. *Chemometrics and Intelligent Laboratory Systems, 74*, 209–213.

Carrasco Moraga, P. A. (2003). Characterisation of the constitution heterogeneity of copper in porphyry copper deposits. Doctoral dissertation, Northern Catholic University in Chile, Faculty of Engineering and Geological Sciences, Department of Geological Sciences.

François-Bongarçon, D. M. (1993). The practice of the sampling of broken ores. *CIM Bulletin, 86*(970), 75–81.

François-Bongarçon, D. (1995). Sampling in the mining industry: Theory and practice, Volume 1: Course notes and transparencies. A short course presented by D. François-Bongarçon in the School of Mining Engineering, University of the Witwatersrand.

François-Bongarçon, D. M. (2002). *MINN596: Sampling in the mining industry, theory and practice*. University of the Witwatersrand, School of Mining Engineering.

Gy, P. M. (1982). *Sampling of particulate materials, theory and practice* (2nd ed.). Amsterdam: Elsevier.

Hahn, R. (2005). *Pierre Simon Laplace, 1749–1827: A determined scientist*. Harvard University Press.

Kahraman, S. A. İ. R. (1999). Rotary and percussive drilling prediction using regression analysis. *International Journal of Rock Mechanics and Mining Sciences, 36*(7), 981–989.

Kumar, C. V., Vardhan, H., Murthy, C. S., & Karmakar, N. C. (2019). Estimating rock properties using sound signal dominant frequencies during diamond core drilling operations. *Journal of Rock Mechanics and Geotechnical Engineering, 11*(4), 850–859.

Minnitt, R. C. A. (2007). Sampling: The impact on costs and decision making. *The Southern African Institute of Mining and Metallurgy Journal, 107*, 451–462.

Minnitt, R. C. A., Rice, P. M., & Spangenberg, I. C. (2007a). Part 1: Understanding the components of the fundamental sampling error: A key to good sampling practice. *The Southern African Institute of Mining and Metallurgy Journal, 107*, 505–511.

Minnitt, R. C. A., Rice, P. M., & Spangenberg, I. C. (2007b). Part 2: Experimental calibration of sampling parameters K and alpha for Gy's formula by the sampling tree method. *The Southern African Institute of Mining and Metallurgy Journal, 107*, 513–518.

Pitard, F. F. (2005). Sampling correctness – A comprehensive guideline. Proceedings of the Second World Conference on Sampling and Blending. Melbourne: The Australian Institute of Mining and Metallurgy.

Pitard, F. F. (2006). MINN596: *Sampling theory and methods*. University of the Witwatersrand, School of Mining Engineering.

THEORETICAL EXERCISE FOR CHAPTER 3

1. What is the purpose of ore sampling? Answer: The purpose of ore sampling is to obtain a representative sample of the ore body for subsequent analysis.
2. What are some of the challenges associated with ore sampling? Answer: Some challenges associated with ore sampling include ore heterogeneity, sample size, and sample preparation.
3. What is the difference between accuracy and precision in sampling? Answer: Accuracy measures how close a sample's estimate is to the actual value, while precision measures the reproducibility of the estimate.
4. What is the difference between a primary and secondary sample in ore sampling? Answer: A primary sample is the initial sample collected directly from the ore body, while a secondary sample is a smaller sample taken from the primary sample for laboratory analysis.
5. Why is it important to use proper sample preparation techniques? Answer: Proper sample preparation ensures that the sample is homogenous and representative of the ore body, and that it will yield accurate and reliable analysis results.
6. What is the difference between grab and composite sampling? Answer: Grab sampling involves taking a single sample from a specific location, while composite sampling involves combining multiple samples from different locations to create a representative sample.
7. What is a sampling error, and how can it be minimised? Answer: A sampling error is the difference between the actual value of a parameter and the estimated value based on the sample. It can be minimised by using appropriate sampling techniques, sample size, and statistical analysis.
8. What are some factors that can influence the quality of an ore sample? Answer: Some factors that can influence the quality of an ore sample include the size and shape of the ore body, mineralogy, and the presence of contaminants.
9. What is the purpose of using duplicate samples in ore sampling? Answer: Duplicate samples are used to verify the accuracy and precision of the primary sample, as well as to identify and quantify any potential sampling errors.
10. What is a sample preparation protocol, and why is it important? Answer: A sample preparation protocol is a standardised procedure for preparing an ore sample for analysis. It is important to ensure that the sample is representative of the ore body and that it yields accurate and reliable analysis results.

CALCULATION EXERCISE FOR CHAPTER 3

1. A bulk sample of 50 kg is taken from a heap of crushed ore with an estimated average grade of 2.5 g/t. If the sample is split into ten subsamples of equal weight, what is the expected range of grades for the subsamples?

   Answer: The expected range of grades for the subsamples can be calculated using the following formula:

$$\text{Range} = 2 \times t \times \left(\frac{\text{standard deviation}}{\text{sqrt}(n)} \right)$$

   where t is the Student's t-value at the desired confidence level (e.g. $t = 2.26$ for 95% confidence), standard deviation is the standard deviation of the population (i.e. the heap of crushed ore), n is the number of subsamples (i.e. 10).

   Assuming a standard deviation of 1.5 g/t for the heap of crushed ore, the expected range of grades for the subsamples is:

$$\text{Range} = 2 \times 2.26 \times \left(\frac{1.5}{\text{sqrt}(10)} \right) = 1.7\,\text{g/t}$$

   Therefore, we would expect the grades of the ten subsamples to range from 0.8 g/t to 4.2 g/t.

2. A cross-belt sampler is used to collect a composite sample of iron ore from a conveyor belt. The sampler is set to collect samples at intervals of 1 m along the belt. If the belt speed is 2 m per second and the cross-belt sampler takes 3 seconds to collect each sample, what is the length of the composite sample collected after 10 minutes?

   Answer: The length of the composite sample collected after 10 minutes can be calculated as follows:

$$\text{Composite sample length} = \text{belt speed} \times \text{sampling time} \times \text{number of samples}$$

   Assuming 60 seconds in a minute, the sampling time for each sample is 3 seconds, and the number of samples collected in 10 minutes is:

$$\text{Number of samples} = (10 \times 60 \text{ seconds}) / (\text{sampling time} + \text{dead time})$$

   where dead time is the time required for the sampler to move to the next sampling position. Assuming a dead time of 2 seconds, the number of samples collected in 10 minutes is:

$$\text{Number of samples} = \frac{(10 \times 60)}{(3 + 2)} = 120$$

Therefore, the length of the composite sample collected is:

$$\text{Composite sample length} = 2 \times 3 \times 120 = 720\,\text{m}$$

3. A mineral deposit has an estimated average grade of 1.5% copper. If a single grab sample of 5 kg is taken from the deposit, what is the expected standard deviation of the sample grade?

   Answer: The expected standard deviation of the sample grade can be calculated using the following formula:

$$\text{Standard deviation} = \text{sqrt}\left(\left(p \times (1-p)\right)/n\right) \times 100$$

   where p is the estimated proportion of the mineral of interest in the deposit (i.e. 0.015 for copper), n is the sample size (i.e. 5 kg).
   Using these values, we get:

$$\text{Standard deviation} = \text{sqrt}\left(\frac{\left(0.015 \times (1-0.015)\right)}{5}\right) \times 100 = 1.9\%$$

   Therefore, we would expect the grade of the 5 kg grab sample to vary by approximately ±1.9%.

4. A mining company wants to estimate the average copper grade of an ore deposit. They collect ten samples from the deposit and obtain the following grades (in % Cu): 1.2, 1.8, 2.1, 1.6, 1.9, 1.4, 1.3, 1.7, 1.5, 2.0. Calculate the sample mean and the sample standard deviation.

   Answer: The sample mean is (1.2 + 1.8 + 2.1 + 1.6 + 1.9 + 1.4 + 1.3 + 1.7 + 1.5 + 2.0)/10 = 1.7%. The sample standard deviation is 0.259%.

   If a sample of ore weighs 500 g and has an specific gravity (SG) of 2.8, what is its volume in cubic centimetres? Answer: The formula for volume is $V = m/\text{SG}$, where m is the mass and SG is the specific gravity. Plugging in the values, we get $V = 500/2.8 = 178.57\,\text{m}^3$.

   A drill core is 2 m long and has a diameter of 5 cm. What is its volume in cubic metres? Answer: The formula for the volume of a cylinder is $V = \pi r^2 h$, where r is the radius (half the diameter), h is the height (length of the core), and π is a constant equal to approximately 3.14. Plugging in the values, we get $V = 3.14 \times (5/2)^2 \times 2 = 39.25\,\text{m}^3$.

5. A blasthole is drilled to a depth of 20 m and has a diameter of 15 cm. If the hole is filled with water up to a depth of 10 m, what is the volume of water in cubic metres?

   Answer: The volume of the blasthole can be calculated using the formula for the volume of a cylinder as in question 6. The height of the cylinder is 20 m and the radius is 7.5 cm (half the diameter). Thus, $V = 3.14 \times 7.5^2 \times 20 = 3536.25\,\text{m}^3$. The volume of water is half the volume of the cylinder since it only fills up to a depth of 10 m. Thus, the volume of water is 3536.25/2 = 1768.13 m³.

6. A bulk sample of ore weighing 500 kg is crushed and split into two equal parts. What is the weight of each split?

   Answer: The weight of each split is 250 kg, since the original sample was split into two equal parts.

7. A pulp sample is diluted by a factor of 10 before analysis. If the assay result is 4.5 g per tonne, what is the original grade of the ore in grams per tonne?

   Answer: To find the original grade, we need to multiply the assay result by the dilution factor. Since the sample was diluted by a factor of 10, the original grade is 4.5 × 10 = 45 g/tonne.

4 Geological considerations in mineral resource estimation

INTRODUCTION

The development of geostatistical techniques facilitated the characterisation and operation of spatial properties by geoscientists and mining engineers. Unlike classical statistics, characterisation considers the spatial dependence of properties and their joint distribution when the data are multivariate. Typical examples include spatial support (the size, geometry, and orientation of the space where observations are made) and data handling principles (Talebi et al., 2022).

In the early foundations of geostatistics, the geological characteristics of the orebody were vastly ignored. This led to the misrepresentation of *in situ* mineral resources and recoverable ore reserves. These challenges gave birth to the statement that "all estimation processes must be guided by the geometry and orebody characteristics." The custodians of the orebody are geologists, either at a mine or at a prospecting camp. In the geostatistical estimation of an orebody, geological characteristics must play a role in determining processes and methods. However, there should be caution in not making the geologist's opinion the only factor.

Consequently, you may ask, if the geologist has no final say in estimation processes, then what are their duties and responsibilities? This question can be answered by examining the general duties and responsibilities of the mining, resource, and exploration geologist and comparing them to those of a mine valuator or geostatistician. Let us consider the role of the geologist working in a mine. Geologists typically examine orebodies and describe the varying characteristics they observe. These materials must be well documented, stored and coordinated in a way that is easy to access. Additionally, it is expected that the shape (geometry) and extent of the orebody will be delimited according to the most recent information available. As part of this process, drilling data, mapping of outcrop and sub-crop exposures, lithofacies and alteration indices, seismic survey and structural data, and ore sample data must all be integrated into a spatially coordinated and readily available database.

The data and information necessary for any interpretation and estimation process must also be readily available. After the above has been accomplished, the geologist must inform the geostatistician of any physical discontinuities (e.g. geological structures, facies, and geometry changes) within the orebody so that these may be included in the mineral resource estimation process. Let us now look at the responsibilities of geostatisticians. The responsible geostatistician must ensure that they

DOI: 10.1201/9781032650388-4

know particular mathematical and statistical prerequisites. The mathematical and statistical criteria for a portion of an orebody must meet specific techno-economic requirements, such as no mixing of different distributions, the portion of the orebody being considered must be quasi-stationary, there are feasible prospects of economic extraction, geological structures and alterations are clearly demarcated, and the data must not be ergodic. You may have noticed that these criteria are primarily mathematical rather than descriptive statistics.

Some geologists may have the necessary scientific training but not the mathematical knowledge to formulate a set of mathematical algorithms to evaluate the orebody and produce a reliable estimate. Thus, geologists and geostatisticians must cooperate closely so that the geostatisticians may guide the geologists regarding how the orebody observations can be used to satisfy the mathematical requirements for estimation. An estimation process will be adversely affected if arbitrary criteria are used either by geologists or geostatisticians. Under most circumstances, this undesirable practice works against effective and accurate mineral resource estimation practices, sometimes leading to mine closure. It would be far more beneficial for geologists and geostatisticians to adopt a cooperative spirit so that the estimation processes incorporate both geological and geostatistical factors. From the above statement, the geological characteristics of the orebody are extremely important, provided they are readily available and utilised in a manner that correctly interfaces with the mathematical rigidity of estimation algorithms.

When estimating mineral resources, geological information is crucial to understanding the distribution and quality of mineralisation. The following are some of the geological information that should be considered for resource estimation:

- Alteration and weathering profiles: Alteration and weathering profiles provide information about the chemical and physical changes that have occurred in the rocks over time. They can indicate the extent of mineralisation and the type of mineralisation present.
- Chip samples: Chip samples are small rock fragments collected from the surface of an outcrop or from a trench. These samples provide a quick and inexpensive way to gather geological information about an area and can be used for mineral or element concentration assays. They can be used to identify mineralisation, lithology, and alteration.
- Drill core logs: Drill core logs provide detailed information about the geology of an area, including the thickness and depth of mineralisation. They also provide information about lithology, ore grade, alteration, and structural features. This information is obtained by analysing the core samples obtained from drilling.
- Geochemical logs: Geochemical logs provide information about the chemical composition of the rocks, including the presence of mineralisation. They can be used to identify mineralisation zones and to determine the grade of mineralisation.
- Geological mapping: Geological mapping involves the systematic collection and analysis of geological data to create a detailed map of the area being

studied. This can include information about lithology, structure, alteration, and mineralisation.

- Geological structures: Geological structures, such as faults and folds, can influence the distribution and quality of mineralisation. Understanding these structures is essential for accurate resource estimation.
- Geophysical data: Geophysical data, including density, gravity, magnetic, seismic, and tomography data, can be used to identify the location and extent of mineralisation. These methods work by measuring the physical properties of the rocks and can provide information about their structure and composition.
- Lithofacies logs: Lithofacies logs provide information about the different types of rock present in an area, including their texture and composition. This information can be used to identify mineralisation zones and to determine the grade of mineralisation.
- Mineralogical information: Mineralogical information provides details about the type of minerals present in the rocks and their distribution. This information is important for understanding the mineralisation potential of an area.
- Mineralisation trends or ore shoots: Mineralisation trends or ore shoots refer to the spatial distribution of mineralisation in an area. Understanding these trends can help to identify areas with high mineralisation potential.
- Orebody geometry: Orebody geometry refers to the shape and extent of mineralisation in an area. An accurate understanding of geometry is essential for calculating the volume and grade of the mineralisation.
- Physical and geotechnical rock properties: Physical and geotechnical rock properties include characteristics such as strength, porosity, and permeability. Understanding these properties is important for assessing the viability of mining operations and determining the most appropriate mining method.

TECHNIQUES FOR ASSESSING GEOLOGICAL UNCERTAINTY

a. Probabilistic modelling

Probabilistic modelling involves assigning probabilities to different geological scenarios based on available data and knowledge. This approach allows for the integration of uncertain geological parameters, such as grade variability, lithological boundaries, and structural discontinuities, into resource estimation. By incorporating uncertainty explicitly, probabilistic modelling provides a more comprehensive and realistic representation of the mineral deposit.

b. Geostatistical simulations

Geostatistical simulations utilise stochastic techniques to generate multiple realisations of the subsurface properties, accounting for geological uncertainty. These simulations consider the spatial relationships and patterns of the deposit, capturing the uncertainty in grade distribution and geological boundaries. Multiple realisations enable the generation of multiple resource estimates, which can be used to quantify the uncertainty associated with the deposit.

c. Uncertainty propagation

Uncertainty propagation techniques assess how uncertainties in geological parameters propagate through the resource estimation process. Through quantifying resource estimates to different sources of uncertainty, such as geological interpretations, sampling errors, or geostatistical parameters, uncertainty propagation provides valuable insights into the reliability of resource estimates.

d. Integration of geological understanding

Collaboration between geologists and resource estimators is crucial for a comprehensive and reliable mineral resource estimation. Geologists provide valuable insights into geological complexity, including lithological variations, structural controls, and geological processes. This understanding allows for the identification of geological domains, the delineation of mineralised zones, and the characterisation of geological uncertainties.

e. Statistical analysis and modelling

Resource estimators, often with expertise in statistics and geostatistics, bring their analytical skills to the collaboration. They provide the necessary statistical frameworks and tools to integrate geological data and uncertainty into the estimation process. Statistical analysis ensures the uncertainty is properly captured and propagated, leading to robust and realistic resource estimates.

f. Considerations for mining engineering

Collaboration with mining engineers is essential to incorporate mining considerations into the resource estimation. Resource estimators can adjust the estimation methodology accordingly by understanding the mining methods, constraints, and operational parameters. This integration ensures that the estimated mineral resources align with the mining strategy and aids in optimising the project's economic viability.

g. Benefits of addressing geological uncertainty

Through quantifying and incorporating geological uncertainty into resource estimation, several benefits can be realised:

- Realistic confidence intervals: Including geological uncertainty allows for calculating confidence intervals around the resource estimates, providing a range of potential outcomes.
- Informed decision-making: Reliable resource estimates enable better-informed decision-making processes, reducing investment risks and allowing stakeholders to make more informed decisions regarding resource development, investment, and operational planning.
- Risk management: Accounting for geological uncertainty helps mitigate risks associated with resource estimation. Understanding the range of potential outcomes allows companies to develop contingency plans and assess the impact of uncertainties on project economics.
- Improved project economics: Incorporating geological uncertainty into resource estimation facilitates more accurate mineral resource

estimation of project economics. Realistic confidence intervals help assess the economic viability of mining projects, optimise mine design, and determine appropriate production schedules.

- Regulatory compliance: Many mining jurisdictions require companies to provide reliable and transparent resource estimates for regulatory compliance. Addressing geological uncertainty assist resource estimators to meet regulatory standards and ensure compliance with reporting guidelines.
- Investor confidence: Transparent and well-documented resource estimates, considering geological uncertainty, enhance investor confidence in mining projects. Accurate estimation and disclosure of geological uncertainties provide a clear picture of the project's potential, increasing trust and attracting investment.
- Effective resource management: Geological uncertainty affects resource management strategies, including mine planning, production scheduling, and ore reserve estimation. integrating geological understanding with statistical analysis and mining considerations helps resource estimators to optimise resource utilisation, minimise waste, and improve operational efficiency.

OREBODY MODELLING

In geology, wireframes typically refer to simplified 2D or 3D digital representations of geological features, such as rock layers, faults, or mineral deposits (Dimitrakopoulos, 1998). These wireframes are constructed using data obtained from various sources, including geological maps, boreholes, seismic surveys, and other geological investigations. They help geologists visualise and understand the subsurface geological features and their relationships. On the other hand, 'iso-surfaces' in geology could be a term used in the context of 3D geological modelling or visualisations. In this case, an iso-surface would represent a specific value of a geological property (e.g. rock density, porosity, or mineral concentration) within a 3D volume of the Earth's subsurface. Iso-surfaces are commonly used to represent geological boundaries, layers, or other features that have a constant value of a particular parameter. The steps to create a 3D geological representation involve delineating geological domains, mineralised zones, and major geological features. Sophisticated modelling software and algorithms are employed to create wireframes that accurately capture the shape and complexity of the deposit. Geological wireframes are a product of geological modelling. There are two main types of geological modelling: explicit and implicit. Explicit modelling is a process where geological features such as faults, folds, and lithological boundaries are explicitly modelled by drawing them in a computer program (Vallejo and Dimitrakopoulos, 2019). This type of modelling requires the input of geological knowledge and an understanding of the spatial relationships between geological features. Explicit models are built using a variety of software tools. Implicit modelling, on the other hand, is a process where mathematical algorithms and statistical techniques from a set of data points infer geological features. These data points may include drill hole data, geophysical data, and surface geological information. Implicit models are built using software tools.

TABLE 4.1

Advantages and disadvantages of different types of geological modelling

	Explicit Modelling	Implicit Modelling
Advantages	1. The ability to represent geological structures with greater accuracy. 2. Easier to incorporate geological knowledge and expertise.	1. Faster and more automated process. 2. Can handle large datasets more easily. 3. Can generate multiple realisations or scenarios.
Disadvantages	1. Requires more time and effort to create a model. 2. Can be subjective and dependent on the interpretation of the geologist. 3. Limited ability to incorporate uncertainty.	1. May lack detail and precision in geological features 2. May not incorporate all available geological data. 3. Reliance on mathematical algorithms can obscure geological understanding.

Table 4.1 shows the advantages and disadvantages of each type of geological modelling.

There are various reasons for creating a wireframe of the orebody, some of which are as follows:

a. Obtaining the extent of an envisioned opencast orebody.
b. Obtaining the volumes blasted from an opencast blasting operation.
c. Determining the extent of the available resource in an underground operation. This is achieved by draping the perimeters of the remaining resource to the wireframe. See Chapter 10 for calculating the actual mineral resources available.
d. Demarcating the transition between reefs on a stacked orebody.
e. Providing the mine planner with the indication of where faults and dykes may occur facilitates the planning process and the positioning of any bracket pillars required from rock engineering.

These are but a few of the many uses that a well-constructed wireframe can be utilised for. This, in turn, begs the question as to what can be considered a well-constructed wireframe?

UNDERSTANDING THE DATA BEING UTILISED

A large proportion of a good wireframe is understanding the relative accuracy of the data being utilised (Vallejo and Dimitrakopoulos, 2019). A ranking of data from most accurate to least can be considered as follows:

a. Survey pegs demarcated as being on the reef horizon or along the edge of the orebody in the case of an opencast operation. Here one has the most accurate measure of the three-dimensional positions of the orebody.

b. On-reef development sampling or reef intersections coordinated from fixed pegs.
c. On-reef stope sampling located from fixed survey pegs.
d. Boreholes that have been surveyed as well as the collar position either surveyed or accurately located from a fixed survey position.
e. Geological mapping data.
f. Two- or three-dimensional seismic data. Although this has been placed last, the accuracy may not be of the highest granularity. Here one needs to consider the level of accuracy that the company provides and place the information thereof in the correct place in the ranking. Additionally, if there has been mining in the area of the seismic data, and the calculations have not been updated with the latest interpretation, taking into consideration available high-quality data, one has at least an indication of the actual accuracy of the seismic data.

SOME FACTORS TO LOOK OUT FOR ON A WIREFRAME THAT HAS BEEN CREATED

a. Sudden changes of dip or direction in strike.
b. Data extending above or below the wireframe, thus creating spikes in the wireframe. Potential reasons for this are:
    i.   Up holes captured as down holes or vice versa. Check and fix and then recreate wireframe.
    ii. Incorrect identification of the rock types during logging. Check and fix and then recreate wireframe.
    iii. Incorrect collar coordinates or elevation capture. Check and fix and then recreate wireframe.
    iv. Incorrect dip used on boreholes not surveyed. Once again check and correct.
c. Areas in which the block of ground has not been constrained and hence the wireframe continues across a loss of ground due to a fault or dyke.

CREATING THE WIREFRAME

a. Make sure that all data is available and in the correct format.
b. Ensure that all geological interpretation of structures has been updated with the latest information.
c. Check to make sure all constraining polygons are in the format required for the software to be utilised. For example, polygons need to be clockwise and closed.
d. Ensure that constraining polygons that have no data within them or less than four points have elevations interpreted from adjacent polygons in which the throw is known.
e. Those polygons that have more than four points within them, do not assign elevations to the perimeter of the block in question. This could lead to

mistakes in interpretation or one not taking every three-pointer combination into consideration. Rather utilise the algorithm provided by the programme to interpret the surface and assign elevations to the periphery. This
is not the geologist shirking his/her duty, but rather a judicious application of understanding human limitations and is not to be frowned upon but
rather commended.
f. Separate the polygons into two datasets one in which the elevations have
been assigned to the perimeter as in point (d) and another into those perimeters which one has not assigned elevations to as in point (e). The reason for
this is that most wireframe packages do not cater for combined processing;
one either has to run the process not utilising elevations on the periphery or
use elevations along periphery.
g. Run the two wireframe processes and combine the results into a single
wireframe.

A method that the authors find valuable is to run a wireframe initially unconstrained, not utilising the perimeters at all (Abzalov, 2006). One then considers
this wireframe with all the superimposed perimeters and looks for areas in which
there are no perimeters, yet one observes an abrupt elevation transition. This, under
many circumstances, may well be an indication of faulting that one is not yet aware
of and can explore and take into consideration. Unbelievably, it has been in the
authors' experience that when this process was requested, on more than one occa-
sion, the geologist in charge suggested that it was a waste of time and that the
department had provided a perfect wireframe, and, furthermore, that it was an
insult to their work ethic. However, this attitude was proved untrue on multiple
occasions, and the observed transitions were later shown to be faulting that had not
previously been picked up.

CREATING A GEOLOGICAL CONFIDENCE MODEL

Basic assumptions on any mine are (Savage et al., 2013):

a. Different areas of structural complexity.
b. These areas are of a definable nature.
c. Within these areas there are sub-areas in which we have a high degree of
confidence in the structural interpretation.
d. One has a well-defined protocol on what is considered negotiable faulting
and what requires re-establishment.
e. If we have a three-dimensional seismic survey, we have a well-defined idea
of what the level of accuracy there is inherent in the survey.
f. The portion of the orebody that we have knowledge about bears some
resemblance to the rest of the area of interest.
g. In order to determine some level of measure of increasing confidence, the
rules applied need to be of a consistent nature across the entire orebody and
also from year to year.

INFERENCES FROM BASIC ASSUMPTIONS

a. If faulting is negotiable, then we can safely ignore that, as actual survey measurements obtain a rigorous estimate thereof.
b. If we have a three-dimensional seismic survey faults greater than the detection limit can be accommodated for.
c. Following from points (a) to (b), it is the faulting greater than negotiable and less than the detection limit that contribute to our uncertainty.
d. If we have no three-dimensional seismic data, then structure is defined from drilling, survey measurements, survey pegs, and extrapolation; hence, faults greater than negotiable that contribute to our uncertainty.
e. There is a relationship between the amount of data we have within a certain area and the confidence we have in our interpretation (e.g. material closer to data equals great confidence).

WHAT DO WE NEED TO DO?

a. Separate the mine into a few areas in which the structural complexity is of a similar nature (e.g. structural domains).
b. Within each structural domain, demarcate and examine the amount and extent of faulting within an area in which the confidence is of the highest calibre.
c. Now utilise a polygon of similar size draped across an area of concern; within this polygon count, the amount of relevant structural features.
d. Compare this figure to that obtained from the well-identified area and calculate the percentage of known structural features.
e. This percentage is then utilised to determine the level of confidence in that particular block.
f. Include as part of review.
g. Identify the block with a polygon and associated percentage within your software and pass it onto the person responsible for the overall confidence model.

The percentage of structural complexity then determines the confidence limits accounted for by utilising something like indicated in Table 4.2.

BASIC ASSUMPTIONS ON OREBODY CONTINUITY

On any mine:

a. The portion of the ore body that we have knowledge about bears some resemblance to the rest of the area of interest.
b. The level of confidence is defined by the proximity to known information.
c. To determine some level of measure of increasing confidence, the rules need to be of a consistent nature.

TABLE 4.2

Example of geological structure confidence intervals

Confidence	Range
Measured	65–99.9% of geological structure identified
Indicated 1	55–65% of geological structure identified
Indicated 2	45–55% of geological structure identified
Indicated 3	35–45% of geological structure identified
Inferred	Less than 35% of structure identified

d. Rules of classification need to pertain to the nature of which we are attempting to classify.
e. Classification is defined by level of confidence and not size of area under consideration.

INFERENCES FROM BASIC ASSUMPTIONS

a. If we have a developed reef in a known area, we can assume the reef extends to *some extent* beyond our area of known information.
b. The extent to which we extrapolate our continuity is defined by *a priori* information.
c. We utilise the same rules on a year-by-year basis.
d. The rule we apply must pertain to levels of continuity.
e. All areas of the mine must be classified.

SO, WHAT DO YOU NEED TO DO?

a. Create an empty block model across the entire mine at a size equal to the selective mining unit (SMU).
b. For each SMU-sized block on the mine.
c. Determine the closest three data points.
d. Calculate the average distance from the SMU-sized block to the known data.
e. Compare this figure to the *a priori* defined levels of confidence.
f. Assign to the block the level of confidence associated with that distance.
g. Include as part of review.

The *a priori* confidence limits can be defined as something similar to Table 4.3. The average distance determines the confidence limits to the three closest neighbours.

Combine the structural and continuity models into a single model taking the lowest confidence as being the indication of confidence. Pass this across to the person responsible for the mineral resource estimation confidence model, so that they can include the final geological confidence model into the estimation confidence model utilising the same approach.

TABLE 4.3

Example of orebody continuity confidence intervals

Confidence	Range
Measured	0–250 m
Indicated 1	250–500 m
Indicated 2	500–750 m
Indicated 3	750–1,000 m
Inferred	Greater than 1,000 m

CONCLUSION

In this chapter, we highlight the critical role of geological characteristics in accurate mineral resource estimation. In the early stages of geostatistics, the significance of orebody geology knowledge was often overlooked, leading to inaccuracies in representing mineral resources and recoverable mineral reserves. Recognising the importance of geological factors, it was established that all estimation processes must be guided by the geometry and orebody characteristics. Geologists, being the custodians of the orebody, play a key role in the geostatistical estimation process. They meticulously examine and describe the orebodies, documenting and coordinating their findings to build accessible databases. The shape and extent of the orebody are delineated based on the latest information, incorporating various data sources such as drilling data, outcrop mapping, lithofacies and alteration indices, seismic surveys, and ore sample data. Geostatisticians, in turn, must possess specific mathematical and statistical prerequisites to ensure that the estimation process aligns with techno-economic requirements. Collaboration between geologists and geostatisticians is crucial to optimally use orebody observations to meet mathematical requirements for estimation, ensuring that arbitrary criteria do not adversely affect the process. Various geological factors, such as alteration and weathering profiles, chip samples, drill core logs, geochemical logs, geological mapping, structures, geophysical data, mineralogical information, mineralisation trends, orebody geometry, and physical and geotechnical rock properties, contribute to understanding the distribution and quality of mineralisation. The chapter further explored techniques for assessing geological uncertainty, such as probabilistic modelling, geostatistical simulations, uncertainty propagation, and integration of geological understanding. These methods allow for the quantification and incorporation of geological uncertainty into resource estimation, providing realistic confidence intervals and aiding informed decision-making, risk management, and project optimisation. To develop accurate orebody models, geological wireframes and iso-surfaces are constructed using explicit and implicit geological modelling. Explicit modelling involves drawing geological features explicitly using geological knowledge, while implicit modelling uses mathematical algorithms and statistical techniques to infer geological features from data points. Each method has its advantages and disadvantages, but a collaborative approach between geologists and geostatisticians is essential for comprehensive and reliable mineral resource estimation. In conclusion, the chapter highlights the indispensable role of geological considerations in mineral resource estimation. By integrating geological understanding with statistical analysis and mining considerations, geologists and geostatisticians can produce robust and realistic resource estimates, facilitating effective resource management, risk mitigation, and informed decision-making in the mining industry.

REFERENCES

Abzalov, M. Z. (2006). Localised uniform conditioning (LUC): A new approach for direct modelling of small blocks. *Mathematical Geology, 38*, 393–411.
Dimitrakopoulos, R. (1998). Conditional simulation algorithms for modelling orebody uncertainty in open pit optimisation. *International Journal of Surface Mining, Reclamation and Environment, 12*(4), 173–179.
Savage, N., Nicholas, L., Wilson, A., & Seery, J. (2013, August). Visual communication of geological confidence–A move toward a less subjective approach. Iron Ore Conference, Perth.
Talebi, H., Peeters, L. J., Otto, A., & Tolosana-Delgado, R. (2022). A truly spatial random forests algorithm for geoscience data analysis and modelling. *Mathematical Geosciences, 54*, 1–22.
Vallejo, M. N. & Dimitrakopoulos, R. (2019). Stochastic orebody modelling and stochastic long-term production scheduling at the KéMag iron ore deposit, Quebec, Canada. *International Journal of Mining, Reclamation and Environment, 33*(7), 462–479.

QUESTIONS FOR CHAPTER 4

1. What are the key geological factors that need to be considered when estimating mineral resources?

    Answer: Geological factors that need to be considered when estimating mineral resources include mineralisation style, geology and structure of the deposit, alteration, and mineralogy.

2. What is mineralisation style, and why is it important in mineral resource estimation?

    Answer: Mineralisation style refers to the way in which the minerals are deposited in the ore body. It is important in mineral resource estimation because it affects the grade and continuity of the deposit.

3. How does the geology and structure of the deposit impact mineral resource estimation?

    Answer: The geology and structure of the deposit impact mineral resource estimation because they determine the shape and size of the deposit, as well as the orientation and distribution of the mineralisation.

4. Why is alteration important in mineral resource estimation?

    Answer: Alteration is important in mineral resource estimation because it can affect the mineralogy of the deposit, which can in turn impact the grade and recoverability of the minerals.

5. What role does mineralogy play in mineral resource estimation?

    Answer: Mineralogy is important in mineral resource estimation because it affects the properties of the minerals, such as their hardness, density, and response to processing, which can impact the economic viability of the deposit.

6. How does the sampling method impact mineral resource estimation?

   Answer: The sampling method can impact mineral resource estimation because it affects the representativeness and accuracy of the data used to estimate the resource.

7. Why is drilling important in mineral resource estimation?

   Answer: Drilling is important in mineral resource estimation because it provides detailed information about the geology, structure, and mineralisation of the deposit, which is necessary for accurate resource estimation.

8. How does the quality of the drilling data impact mineral resource estimation?

   Answer: The quality of the drilling data impacts mineral resource estimation because it affects the accuracy and reliability of the resource estimate.

9. What is the role of geostatistics in mineral resource estimation?

   Answer: Geostatistics is a statistical approach that is used to analyse and interpret geological data. It is used in mineral resource estimation to provide a more accurate and reliable estimate of the resource.

10. Why is it important to consider the uncertainty and variability of the geological data in mineral resource estimation?

    Answer: It is important to consider the uncertainty and variability of the geological data in mineral resource estimation because it can impact the accuracy and reliability of the resource estimate and can also affect the economic viability of the deposit.

5 Preliminary mineral resources estimation processes

INTRODUCTION

Mineral resource estimation, a cornerstone of successful mining endeavours, serves as the key to unlocking the hidden treasures within the Earth's crust. Yet, this pursuit is not without its challenges, as it demands a thorough understanding of local data, geological variables, and the intricate patterns that govern their spatial correlation. In this journey of exploration, we embark on an enlightening odyssey, unravelling the preparatory steps that lay the groundwork for precise and reliable mineral resource estimation. At the heart of this endeavour lies an array of local information, each with its unique scale and accuracy. Direct measurements, such as core data, allow us to analyse and interpret subsurface samples, revealing significant insights into the distribution of elements such as copper and cobalt ore grades. On the other hand, indirect measurements tap into rock responses without physically extracting samples from the Earth, encompassing techniques such as drillhole logs and seismic attributes. Complementing these local data sources is the wealth of analogue conceptual information acquired from sources external to the deposit, such as mature deposit outcrops, flume experiments, and physical-based models. Integrating this vast array of information is the greatest challenge and strength of geostatistical reservoir modelling. Uniting the pieces of the geological puzzle, decision-makers must determine the stationarity of the data, assessing whether the statistical properties remain consistent across different deposit regions. Furthermore, a crucial aspect involves establishing spatial correlation patterns and understanding how neighbouring data points influence each other and affect the estimations away from the sampled locations.

A series of preparatory steps come into play to achieve the most accurate estimations. Exploratory data analysis (EDA) unveils the hidden trends and patterns within the data, shedding light on the spatial distribution of crucial variables. By analysing the distribution types of grade and geological variables, experts select appropriate estimation models that best capture the nature of the deposit. Estimation domains serve as spatially constrained areas, each characterised by its unique geological signature. These domains allow tailored estimation methodologies to be applied, maximising precision and accuracy across the deposit's diverse regions. Variography emerges as a geostatistical wizardry, quantifying geological attributes' spatial variability and continuity. This powerful tool reveals the underlying spatial correlation structure, providing invaluable insights for fine-tuning estimation models and ensuring robust predictions.

DOI: 10.1201/9781032650388-5

Pre-estimation cross-validation paves the way for rigorous testing and optimisation, verifying the performance of estimation techniques before delving into full-scale resource estimation. This step guards against overfitting or underfitting, ensuring that the chosen methodology aligns seamlessly with the data at hand. Finally, the compass of kriging neighbourhood analysis guides us in selecting the optimal arrangement of neighbouring data points to enhance the accuracy of estimations. This delicate balance between local precision and global consistency ensures the creation of a coherent and comprehensive mineral resource model. As we embark on this enlightening journey through the preparatory steps in mineral resource estimation, we uncover the intricate tapestry of geological insights and geostatistical techniques that govern the assessment of Earth's hidden riches. Join us in this exploration of discovery as we illuminate the path toward robust and reliable mineral resource estimations, driving the mining industry towards a future of sustainable prosperity and resourceful innovation.

THE PREPARATION PROCESS FOR MINERAL RESOURCES ESTIMATION

To ensure the estimation of metal distribution within a deposit accurately reflects reality, it is important to follow a meticulous process. This chapter outlines the key steps involved in preparing for, or updating, a mineral resource estimate. While adhering to these processes, geostatisticians and resource geologists can enhance the reliability of their assessments. The processes are as follows in order of priority:

- Validate and prepare drill hole information in the correct electronic format.
- Validate and prepare geological mapping information in three-dimensional space and electronic format.
- Examine sampling data daily.
- Prepare lithological and alteration interpretations.
- Review prior life-of-mine plans.
- Review and validate the current structural interpretation.

VALIDATE AND PREPARE DRILL HOLE INFORMATION ACCURATELY AND IN THE CORRECT ELECTRONIC FORMAT

The first step in the estimation process is to validate and prepare the drill hole information in the appropriate electronic format. This involves verifying the integrity of the data, checking for any inconsistencies or errors, and ensuring that it is compatible with the subsequent analysis. Accurate and well-formatted drill hole data form the foundation for reliable mineral resource estimation. Depending on the geological and estimation software being used, several input files may be required, these are generally categorised as minimum files required, and additional files.

Minimum files required:

- Collars (this file contains the borehole collar coordinates and is generally in the form Borehole number (BHID), X (Borehole X collar), Y (Borehole Y collar), and Z (Borehole Z collar)). Note that a borehole collar refers to the top or starting point of a borehole.

- Borehole Survey file (referred to as the Survey file and is generally in the form of BHID, survey depth and the hole).
- Assay (containing the assay values, in the form BHID, Sample Number, From, To, and the required assay values).
- Lithology (containing geological, lithology information in the format BHID, From, To, Rock type, Lithology, Stratigraphy, etc.).

Additional files:

- Weathering (in the Format: BHID, From, To, Weathering information).
- Mineralisation (BHID, From, To, Mineralisation).

When validating the data that is to be used, the following minimum checks are recommended:

a. Collar coordinates (x, y, z).
b. Borehole surveys.
c. Assay QA/QC analysis is completed, and the assay data is found to be free of bias.
d. Assay values captured and validated against the original assay certificates.
e. Geological logging is captured along with prescribed attributes, for example, lithology, alteration, and weathering, The validation check should include a peer check and sign off.
f. Plot the drill hole data in 3D in the modelling software and visually check that the drill holes plot correctly, for example, that the collar positions align with the locations that they were drilled from and that the drill hole orientation is correct (the Borehole surveys are correct), be it from topographic projection or underground excavations.
g. Complete a peer-review check and formally hand over the data set.

The validation and hand over of drill hole data is normally undertaken on an annual basis. It is good practice to provide a summary table of the new drill hole data (as well as documenting any revisions or edits that have been made) and make it available to the persons undertaking the geological model update and the estimation process.

If any drill holes are excluded from the data set, the reason for exclusion should be documented; the data should NOT be deleted from the master database, only from working databases if necessary. However, the 'questionable' data should be flagged in the master database.

A word of caution is to NEVER work in or edit the master database. Always use a working copy, so that if anything goes wrong a backup is available. One of the authors is aware of a company who corrupted their master data base and had no back up. All the data had to be recaptured at a significant cost and effort.

VALIDATE AND PREPARE ANY MAPPING INFORMATION IN THREE-DIMENSIONAL SPACE AND ELECTRONIC FORMAT

Next, geostatisticians need to validate and prepare the geological mapping information in three-dimensional space and electronic format. This ensures that the geological context of the deposit is properly captured, enabling a comprehensive understanding of its spatial characteristics. Like the drill hole data, the geological mapping information must be validated for accuracy and formatted for compatibility with subsequent analysis. The estimate that is prepared is based on the geological interpretation of the deposit, which is based on geological mapping and observations. Accurately recording a modelling of structural geological data (e.g. faults and dykes) as well as lithological data is critical for a valid geological; interpretation to be achieved. The following aspects should be considered:

a. Ensure that all mapping is up to date, including bench, face, or development (e.g. flat back and sidewall) mapping. Ensure that the mapping data is correctly georeferenced in 3D space and is in an electronic format that can be readily ported between various modelling software programs, typical file formats are .dm, .shp, .dxf, or .dwg; however, always ensure that data are not lost when transferring files between different software's or file formats.
b. Accurate survey data are critical to determining the location of excavations and the relative locations of the data that are to be used in the estimation process. Ensure that the coordinate system is consistent between different data sources.

EXAMINE SAMPLING DATA DAILY

To maintain data quality and address any potential issues promptly, a diligent examination of sampling data daily is essential. This regular scrutiny helps identify outliers, assess the representativeness of the samples, and identify any potential biases or irregularities that may impact the estimation process. By staying vigilant and proactive in data monitoring, geostatisticians, and resource geologists can maintain the integrity of the estimation process. In most mines, there is a continuous flow of new samples going to the assay laboratory and the resultant metal assays returning to the relevant department for attention. Depending on the operational setup, this could be the Geology Department, the mineral resource or Ore Reserve Department, the Grade Control Department, or a stand-alone Sampling Department. The locus of control and operational structure is in part determined by how the sampling data are used in the estimation process. Given the volume of data, it is good practice to verify and validate any updates that are made to the sampling database daily.

It is recommended that the sampling data be stored in a different database to the geological drilling data. In most operations the sampling data are used for grade control purposes (e.g. blast hole sampling, channel sampling or vertical stratigraphic sections or rock chip samples); however, on Witwatersrand gold mines in South Africa, as an example, it is standard practice to combine assay data sourced from geological drilling as well as routine 'grade control' sampling.

It is axiomatic that the use of the most recent, accurate, sampling data will provide the 'best' estimate of the metal grade and content of the rock that you intend to mine. The preceding statement is oversimplistic, given the multitude of variables and decisions that must be considered when undertaking a mineral resource estimate, the point being, it is naive to expect that you will get an accurate assessment of your planned mining areas based on data that are six months behind where mining is currently being undertaken.

PREPARE ANY FACIES INTERPRETATION

Preparation of lithological and alteration interpretations is a critical step in understanding the geological complexity of the deposit. Geostatisticians and resource geologists must carefully analyse the available data and establish accurate lithological and alteration models. These interpretations provide valuable insights into the spatial distribution of different rock types and mineralogical variations, enabling more precise estimation of metal distribution. We know that there is often a strong relationship between metal grade and rock type or metal grade and another measurable feature of the host rock, for example, high metal grades may be associated with a specific type of alteration or sulphide group of minerals. There is need to ensure that these observations are accurately recorded and used in geological interpretations.

This process would include a review of, but is not limited to:

a. Changes in the interpretation of the location and characteristics of the metal bearing horizon.
b. Any areas where the mineralised horizon pinches out. In narrow tabular deposits, this is often referred to as waste on contact.
c. Areas of the deposit that have specific, identifiable, geological characteristic, such as wash-out zones and overbank areas, that is, geological facies types.
d. Any areas where the geological facies or litho-type appear similar, but, however, have distinct grade differences.

These observations must be documented and stored for reference purposes and to assist with first order geological interpretations. These documents should include why and how the area was derived. Additionally, all areas must be demarcated using polygons that have a field that identifies the polygon and the geozone that it is associated with. When completing the polygons, ensure that the polygons are closed on the correct elevation(s) to avoid overlap or gaps with other polygons, and to prevent malformed strings.

REVIEW PRIOR LIFE-OF-MINE PLANS

Reviewing and analysing prior life-of-mine plans contributes to a comprehensive understanding of the deposit's history and extraction patterns. This knowledge helps in assessing any potential biases or limitations in previous estimations and guides the current estimation process. A thorough mineral resource estimation of past plans aids in minimising uncertainties and improving the reliability of the updated mineral resource estimate. When embarking on the estimation process, reviewing the previous year's mine plan and schedule is critical, as the estimate is the basis for mine planning, and determining the amount of money the mine could make.

When undertaking this review, consider the following:

a. Areas within a mine where reclamation or reclamation mining is occurring. These areas should be demarcated as specific mining districts in order that the correct mining availability classification can be applied.
b. Project areas such as new incline or decline shafts and/or below infrastructure mining such as winze mining must be identified as a specific mining area and be demarcated via specific polygons.
c. Any areas outside of the mine's infrastructure that cannot be mined must be identified and documented.
d. Any areas within the mine's infrastructural limits that have been abandoned to be identified and demarcated.
e. Any changes to mining rights need to be identified and documented.
f. Major structural features that are known to have lateral movement, or are syn-depositional, must be identified and demarcated for use as hard boundaries.
g. Examine areas where the previous estimate did not meet expectations (both in terms of lower as well as higher than planned metal grades or thicknesses):
    i. Are there any structural features which should have been used as hard boundaries but were not identified or demarcated?
    ii. Are any data points (e.g. drill holes or other sampling data) incorrectly positioned (poor or invalid surveying),
    iii. Were the data points assigning the correct estimation domain designator, or have the data points been assigned incorrect assay values or thickness? The latter scenario is often encountered when using historical databases that have not been subjected to the same level of validation as is expected in terms of current industry-accepted practices, and
    iv. Were all the geozones and other demarcation polygons used?
h. The review and validation process must be completed prior to commencing a mineral resource estimate.

DO A DETAILED EXAMINATION OF STRUCTURAL INTERPRETATIONS

The current structural interpretation plays a vital role in understanding the structural controls of the deposit and its influence on metal distribution. Geostatisticians and resource geologists need to review and validate the structural interpretation, ensuring its consistency with the available geological and geophysical data. A reliable structural interpretation forms the basis for accurately characterising the spatial continuity and variability of mineralisation within the deposit. Note, there are overlaps between these validation checks and those identified in the preceding action step:

a. Did the structural model utilise all of the data obtained during the previous year?
b. Does the structural model honour the data?
c. Is the 3D model a correct representation of the structural interpretation?
d. Where there is limited data available, does the structural interpretation makes sense?
e. If sub-crops are present, ensure that they do not overlap with overlying horizons.
f. All changes must be formally documented.

Following the validation process, modelling may be commenced, using the following process flow.

MODELLING PROCESS FLOW

It is important to note that this is an overview of the process, and that further detail is provided in the text. In all cases:

- Record and document your work in a clear and concise manner. Make certain there is sufficient detail that a third party could take your notes and reproduce your work.
- Even though it is not specifically noted, at each step of the process summarised below, it is expected that you will document the steps you followed, results achieved, and errors and irregularities dealt with.
- Ensure that diagrams are clear and legible.
- If you have specific legal requirements or company protocols to follow, ensure that the relevant documentation, sign offs, and hand over documents are completed and correctly filed.

PART 1 – EXPLORATORY DATA ANALYSIS[1]

1. Ensure that you have the latest geozone (geological and geostatistical criteria combined) polygons and that they are in good order, ensuring no gaps between zones.
2. Make sure that you have the latest mining cut polygons if areas require different mining cuts or benches depending on where they are in the mine.
3. Import sampling data.
4. Check sampling data. Yes, we are aware that at this time, all problems should have been taken care of; however, inevitably, there will be glitches and errors that, even at this late stage, will need to be fixed. Fix any problems that may be remaining.
5. Import drill holes.
6. Check drill holes and report any holes that were not imported to the responsible geologist. Once again, we are aware that all problems should have been taken care of at this time!
7. If dealing with 3-parameter log-normal distributions, calculate third parameters per geozone.

8. Determine if there is a relationship between the planned mining height and the mineralised horizon thickness. If a relationship exists, model this, ensuring you consider varying areas of hanging wall and footwall physical properties.
9. Note differences in 7 and 8 and reconcile against last year's modelling.
10. Place modelling in the report and comment on exceptions.
11. Do general statistics on point data to assess distribution and correlation within and between datasets.
12. Capture this into the report and make comments on year-on-year changes. You will use this later for some of your reconciliation explanations.
13. Run outlier analysis on point data on a domain-by-domain basis, putting general outliers treated results report into a modelling report.
14. Do general statistics on a domain-by-domain basis point data after outlier analysis.
15. Do trend analysis on a domain-by domain and place into report noting any changes from the previous year and observing any significant trends.
16. If modelling a tabular orebody, doing channel width versus cumulative value sum plots and checking for signs of incorrect domaining is important and incorporating results into the report.
17. Do variance analysis, comment on exceptions, or tweak zonal boundaries and place them into the report.
18. If tweaking zonal boundaries are required, redo points 1–17 for geozones affected and place them into the report.
19. Update the database to reflect the latest third parameters.
20. Print out the Sarbanes–Oxley Act document for processes completed and sign off.

Part 2 – Regularisation and support size

21. Regularise the sampling data into blocks of various sizes per domain – this depends on data density. A good starting point is to commence at half of the average sampling spacing and then double the size until the entire domain is covered by a single block.
22. Run stats on data, examine for inconsistencies, and place into the report with comments if necessary.
23. Calculate in-block variance versus the number of samples for each block size. Use this to determine how many samples are required for a certain size block in order to achieve a representative mean. Document analysis.
24. Now utilise the number of samples required to determine what percentage of the blocks have representative means – document analysis in the report.
25. Utilise point number 24 to determine the number of samples cut-off for further analysis and the associated error with utilising the reduced number of samples. Document the results of the analysis in table form in the report document.
26. Model uneconomic blocks reconciled against the previous year and commented on changes in the report document.
27. Print out the Sarbanes–Oxley Act document for processes completed.

PART 3 – DECLUSTERING AND NUGGET EFFECT ANALYSIS

28. Compete a declustering analysis per domain and compare the results to the previous year's exercise. Ensure to comment on changes in the report document.
29. Update your database mean entries to reflect the latest information in the report.
30. Calculate normal score point semi-variograms (note that other practitioners may prefer to work on three-parameter log-normal distribution data).
31. Model normal score or log point semi-variograms. Check to ensure that the overall sill equals the normal score or log variance of your points. Make sure you take cognisance of the number of pairs when modelling. Place modelling with any necessary comments into your report.
32. Calculate and check your nugget-to-sill ratios, examine any suspicious models, and fix them if required. Place a results table with a remark's column into the report document. Note that this is only relevant if the software does not automatically scale the total variance to 1. (Note that an experimental semi-variogram scaled to 1 does not allow for checking of 1.)
33. Verify that there have been no changes to selective mining unit (SMU) sizes. If there have been changes, modify these in the block information table and include a copy in the report with any necessary comments.
34. Make sure the block information table shows the Macro kriging block sizes if one is going to use Macro kriging. If these have changed, modify the file and comment in the report.
35. Calculate the dispersion and estimation variances using the normal score or log point semi-variograms modelled and the block information. Place results into your report.
36. Print out the Sarbanes–Oxley Act document for processes completed.

PART 4 – CALCULATE EXPERIMENTAL VARIOGRAMS AND ASSOCIATED PARAMETERS

37. Calculate your point semi-variograms, that is, semi-variograms using the untransformed assay data (see Section 'Data declustering and support size determination' to calculate experimental semi-variograms)
38. Model the point semi-variograms. Check to ensure the overall sill equals variance of the data utilised. If using a relative semi-variogram, where the variance has been normalised to 1, the sill is equal to the points' variance divided by the mean squared. Make sure you take cognisance of the number of pairs when modelling.
39. Calculate and check your nugget-to-sill ratios, examine any suspicious models, and fix them if required. Place a results table with a 'remarks' column into your report document.
40. Calculate your block semi-variograms (e.g. optimum block size can be calculated using kriging neighbourhood analysis or sample configuration analysis)
41. Model your block semi-variograms. Check to ensure the overall sill equals the block variance or the block variance divided by the mean square if these

are relative semi-variograms. Make sure you take cognisance of the number of pairs when modelling. Place modelling with any necessary comments into your report.

42. Calculate and check your nugget-to-sill ratios, examine any suspicious models, and fix them if required. Place a results table with a 'remarks' column into your report document.

43. Print out the Sarbanes–Oxley Act document for processes completed.

PART 5 – PRE-ESTIMATION CROSS-VALIDATION

44. Do cross-validation exercises for all block sizes. The ultimate goal of cross-validation is to enhance the accuracy and reproducibility of the estimates. Iteratively refining the sample set and evaluating various block sizes help with the identification of the optimal combination that minimises biases and maximises the fidelity of the estimation results. This rigorous approach ensures that the final model provides a true reflection of the metal distribution within the deposit. Remember that you must do a few of these with different maximum sample criteria to obtain a reliable sample set to model. Cross-validation is also important in kriging to minimise conditional bias and smoothing prior to the actual estimation.

45. Model the relationship per geozone between the number of samples and absolute error per block size. Use this to obtain the minimum and maximum samples used in kriging (Figure 5.1).

46. Place results of these in the report, commenting where parameters were borrowed from if a domain had insufficient data to accomplish the step.

47. Model the relationship per zone between kriging regression slope and absolute error. Use this to obtain a minimum kriging regression slope cut-off, below which the ordinary kriging may be replaced by the simple kriging.

48. Place results of these in the report, commenting where parameters were borrowed from if needed.

49. Model the relationship per zone between kriging efficiency and absolute error. Use this to obtain a minimum kriging efficiency cut-off, below which the kriging is to be replaced by the next block size and hence will be discarded for this block size.

50. Place results of these in the report, commenting where parameters were borrowed from if needed.

51. Update the database entries for min and max samples and kriging regression slope and kriging efficiency to reflect what is in the report.

52. Do final regression analysis on results and store data in required files and documents.

53. Print out the Sarbanes–Oxley Act document for processes completed.

PART 6 – LOCAL MEAN DETERMINATION

54. Do local mean search analysis per block size and obtain search radii and minimum samples per block size and per zone.

55. Place results of these in the report, commenting where parameters were borrowed from if needed.
56. Update the database entries to reflect what is in your report.
57. Calculate local mean models for each block size.
58. Examine local mean models for anomalies and areas where you may have to relax the criteria slightly.
59. Place comments with screen grabs into the report.
60. Where you are going to do block kriging, use your point semi-variogram to calculate the discretisation required for kriging. Discretisation is an essential process in geostatistics, involving the subdivision of space into a series of points or the transformation of a coarse grid into conforming fine grids (Manchuk, 2010). Each discretisation point or node represents a volume within the grid block, playing a crucial role in capturing the variations of rock properties during geostatistical modelling. Grid refinement and point modelling are fundamental to accurately represent the spatial variability of rock properties within the target grid. This process allows for a more detailed understanding of the distribution and behaviour of the variable of interest. Similar to concepts found in computational fluid dynamics and other numerical methods, the discretisation of grid blocks considers the small scale of data relative to the block size. Practical applications of block discretisation are particularly significant in geostatistical workflows employed in mining and petroleum industries. These industries rely on techniques such as kriging and simulation to estimate and simulate random variables representing grid blocks. By discretising the blocks using a set of locations, the data-to-block average covariances can be calculated, enabling accurate estimation and simulation of the variable at the block level.
61. Place these results into the report.
62. Print out the Sarbanes–Oxley Act document for processes completed.

PART 7 – MACRO KRIGING CONSIDERATIONS

63. Prepare the sampling data for macro kriging by regularising or rasterising it into a standardised block size, such as 20×20 m blocks. This step can be particularly useful when there are insufficient data available for life-of-mine planning.
64. Conduct a variance analysis on the regularised data to determine the domains for macro kriging. Keep in mind that these domains may differ due to central limit tendencies. Consequently, when using data that has been regularised into larger block sizes, you may need fewer domains.
65. Carefully examine the post-plot of the macro kriging domains to ensure their coherence and logical representation.
66. Document the analysis and findings in the report.
67. Employ local mean search analysis using macro geozones and include the results in the report.
68. Create a local mean model and ensure there are no gaps or anomalies in the data.

69. Repeat the regularisation process for the base size used in macro kriging multiple times, using the different numbers of samples.
70. Use the regularisations from 69 to calculate semi-variograms.
71. Model these semi-variograms to determine the nugget effects for each set of samples. Remember that as the number of samples within a block increases the nugget effect on the semi-variogram should decrease. If this relationship is not observed, revisit the process, identify any issues, and make appropriate adjustments.
72. Use the models in 71 to model the relationship between the number of samples inside a block versus the corresponding nugget effect.
73. Include these models in the report along with relevant comments.
74. Discretize the block kriging using the base semi-variogram model. Present the results, including graphs, in the report.
75. Ensure that the latest information, especially the drilling data, is incorporated for the kriging run.
76. Incorporate the nugget effects per block into the regularised data file using the model obtained in step 72.
77. Set the database search range to a size that is sufficiently large to obtain the required number of simple kriging estimates.
78. Print out Sarbanes-Oxley (SOX) Act document for processes completed.

Part 8 – Kriging choice

Four kriging techniques are typically preferred in this context. These techniques include ordinary kriging, simple kriging, macro kriging, and multiple indicator kriging.

79. The first approach is to Krige the micro blocks for grade control models or mine planning purposes. It is important to apply limit cut-offs to the micro blocks for further analysis. Ensure to account for the mining block factor and the final regression corrections need to be applied.
80. Secondly, the macro blocks should be Kriged as well for mine business plan and life-of-mine plans.
81. In order to create a model of estimation confidences on the macro kriging blocks, the lower confidence limit values and the protocol on confidences are utilised. It is important to note that this step is unnecessary for the micro blocks, as estimates below measured confidence are not allowed to pass through to the final grids.
82. Calculate the geological confidence intervals based on regulatory specifications or deposit type.
83. The micro and macro blocks must be combined into a single model, which closely resembles the size of the SMU. It is crucial to ensure that the micro blocks override the macro blocks, not the other way around.
84. The geology confidences and estimation confidences are then combined into a single model.
85. The final confidence value is determined by selecting the lowest confidence between the geology confidences and the estimation confidences.

86. A thorough examination is conducted to identify and address any bull's eyes, and appropriate overwriting is performed as needed.
87. Graphics depicting the findings and results must be included in the report, which can be utilised later for the resource presentation.
88. At the SMU support size, uneconomic blocks are modelled, updated, and documented.
89. The processes completed should be documented by printing out the Sarbanes–Oxley Act document.

PART 9 – ECONOMIC ASSESSMENT FOR MINE PLANNING

90. Update dispersion variances and uneconomic blocks in both planning and the resource software, get a colleague to check and formally sign off that they have checked.
91. Where macro kriging blocks are in the estimation grid, a change of support is required as the block size exceeds the SMU size. Discuss with the mine planning manager if the planners will use selective mineral resource estimation; no further work is required as the support change and uneconomic blocks form part of the grid utilised by planning; however, if the planners require the percentage extraction and extraction value as fields in the estimation grid, the following procedure must be followed.
    a. Run an export of the grids.
    b. In the micro grids, assign 100% to the extraction value field and the estimated value to the extraction value field.
    c. Take the macro grid export into a percentage extraction calculation software such as a spreadsheet or software that allows for scripting.
    d. Update the uneconomic blocks in the software and have a colleague check and sign off.
    e. Calculate the percentage extraction.
    f. Calculate the extraction value.
    g. Save the file in a format that can be used by planning.
92. Import grids into the mine planning environment or hand over the estimate. Check to ensure that:
    a. Any changes in shaft scenario polygons have been corrected.
    b. Any changes in geozone polygons have been corrected.
    c. The latest geological reef horizon model has been imported, and there are no gaps.
    d. All grid sizes available have been utilised in a composited Kriged grid model.

PART 10 – MODEL HANDOVER CONSIDERATIONS

93. Supply the mine planning and mining personnel with handover SOX documentation.
94. Add a content index page to your report.

Most importantly: You will have noticed the amount of work required for an effective modelling process. Late starting should be accepted. In the remainder of the document, you will discover how to implement the processes and sub-processes required and how the results of each analysis fit into the following process until the estimation and cross-validation steps are completed.

SAMPLE TYPES AND ASSOCIATED UNCERTAINTIES

Although the types of samples are many and varied, there is the subconscious perception that a sample is limited to the result of a physical piece of something obtained and its subsequent assay. Nothing could be further from the truth! Nevertheless, this bias does occur and inevitably results in insufficient attention being provided when sampling other variables that may not conclude in a resulting assay. This is why adequate quality assurance and quality control (QA/QC) should be exercised at all steps along the sampling acquisition and not only for the assay process. One of the tools required to minimise uncertainties is task observation.

The task observation comprises two aspects, the planned task observation (PTO) or the unplanned task observation (UTO); the observations must be done by the supervisor or his second in charge. The number of observations is at the supervisor's discretion, but each person responsible for sampling should be observed at least once every quarter.

THE PLANNED TASK OBSERVATION

This method must be employed at a minimum on an annual basis when an employee returns from a leave of absence, for example, annual leave or a period of absence longer than two weeks. The methodology is as follows:

a. Decide on who is to be audited.
b. Inform the relevant person the day before that he/she will receive a PTO on the following day's work, allowing sufficient time to collect the work plan for the following day.
c. Proceed to the workplace with the person and use the workflow template to assess the incumbent's knowledge base.
d. Continue to follow the workflow in the office until he/she has completed the daily required work.
e. Identify and document shortcomings.
f. Schedule a follow-up meeting in which the shortcomings are brought to the attention of the person in question in a non-threatening environment and attitude.
g. Suggest and document remedial actions to point (e) above.
h. Have the person sign the document in point (f) above and provide a copy.
i. Schedule a follow-up to ensure any shortcomings have been taken cognisance of.

This ensures quality assurance as it determines whether the person sampling knows what is supposed to be done.

THE UNPLANNED TASK OBSERVATION

Unplanned task observations may not always be legally undertaken in some districts or companies. Before considering this option, determine if you can undertake this activity. This method is employed on a cross-section of the department, with the provision that all persons are assessed at least twice annually. The methodology is as follows:

a. Decide on who is to be audited.
b. Do not inform the person that they are going to be checked.
c. Allow the person to precede the auditor to the workplace.
d. The auditor follows the person at later stage and proceeds to the working place at a leisurely pace to allow the person time to have started work by the time the auditor arrives.
e. Upon arrival, the auditor requests the person's field notes and examines the work done and the corresponding field note entries to ensure that the work done was of sufficient diligence. The following provides examples of checks.
   i. Check that the whole sample has been taken, in the case of chip sampling.
   ii. Check that correlation between samples has been correctly documented and that all faults, dykes, etc., have been ascertained. In the case of chip sampling.
   iii. Check that geological observations are correct and have been neatly documented. Both chip sampling and core logging
   iv. Observe the general neatness and legibility of field book observations.
   v. Check that differences and rock types have been correctly identified.
   vi. Check that all widths have been measured and documented correctly.
   vii. Check that samples have been chipped to an even depth and that all of the sample has been chipped. In the case of chip sampling.
   viii. Check that cutting of the core was done at the correct positions.
   ix. In the case of halving the core check that cutting insofar as possible is uniform and of a good quality.
   x. Check that all samples have been bagged and tagged and that bags have been securely tied to prevent spillage and cross-contamination.
f. From this point onwards, the audit takes the form of a PTO.
g. The surface part of the unplanned observation is scheduled for another day and follows a checking process of surface and office work once completed.

Task observations are not only limited to those taking the samples. However, they should also be done regularly in the assay laboratories by competent persons who understand the required process. Once again, reporting back should be done with a non-threatening demeanour to ensure compliance. Schedule a follow-up on the laboratory to check that any suggestions have been considered.

THE QUALITY CONTROL AND QUALITY ASSURANCE GRAPHS

Accuracy, precision, and bias are statistical concepts that play a significant role in characterising the quality of data. Accuracy refers to the degree of proximity between a measured value and a known value. Precision, on the other hand, quantifies the reproducibility of a measurement, indicating how closely repeated measurements of the same quantity align with each other. Bias, the third term, measures the extent to which the outcome of an analysis deviates from certified or expected results.

In the context of sample assay repeatability, an effective representation can be achieved through the calculation of the coefficient of variation. This metric is derived from the assessment of differences between original samples and their duplicates (Abzalov, 2009; Stanley and Lawie, 2007). Abzalov (2011) proposed specific levels of precision for different types of deposits based on the coefficient of variation. While the coefficient of variation is a widely utilised relative index of variation, there are other indices available, such as the Reduced Major Axis suggested by Sinclair and Bentzen (1998). Geoscientists commonly employ this model to identify bias in paired data (Sinclair and Blackwell, 2006). In addition, Abzalov (2008) introduced the Relative Different Plot as a graphical tool for assessing the factors influencing precision errors. This tool offers valuable insights into understanding the sources and magnitude of precision-related discrepancies in data analysis. A variety of statistical and graphical methodologies can be used to ensure that the assay results from the laboratory are fit for purpose to be used in an estimate. We will discuss some of these graphs in subsequent sections.

Check sampling control graph

This method offers valuable insights not only for task observation purposes but also serves as an invaluable tool in identifying issues within the assay laboratory. The methodology outlined below should be followed:

a. Identify a small number of highly reliable samplers and instruct them to collect check samples in immediate proximity to the existing samples. Record the pairs of sample and check sample for each reef and homogeneous area.
b. Develop a comprehensive database comprising a minimum of 1,000 pairs for each area mentioned in point (a).
c. Utilise the database to calculate the variance between the sample and check sample for each respective area.
d. Employ the variances obtained in point (c) to determine upper and lower confidence limits at 90% and 95% levels.
e. Regularly update the database at least twice a year to ensure the data remains current.
f. Generate a graph depicting the sample/check sample variances for each individual and compare them to the established standard (refer to Figure 5.1).

If an individual consistently deviates from the established tram lines (e.g. the upper and lower confidence limits) when conducting their work, it suggests that they might be dividing a single sample into two. Such behaviour indicates a lack of due diligence

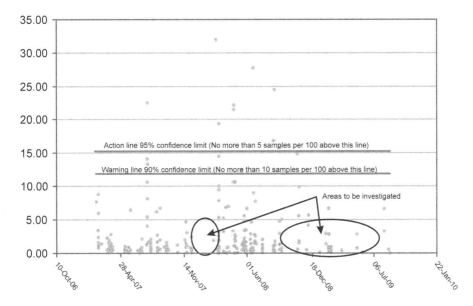

FIGURE 5.1 Example of check sampling control graph.

on the part of the individual, necessitating an UTO. It is important to inform the person in question about the observed deviation. Alternatively, if there are numerous instances of pairs falling outside the tram lines, a similar course of action should be taken. This suggests that there may be a systematic issue or inconsistency in the process that needs to be addressed. In a different scenario, if all individuals consistently display variances that are either well within the tram lines or excessively outside them, it is prudent to consider evaluating the assay laboratory. Scheduling a visit to the laboratory, accompanied by a task observation, can provide valuable insights into the laboratory process. This assessment will help identify any potential factors contributing to the observed variations.

The HARD (half absolute relative difference) graph

HARD, as stated by Shaw (1997) is a precision measurement tool, hence, precisions measured from the use of HARD are comparable from one deposit to the other. Ideally, about 90% of results should be within 10% of the original value as shown on Figure 5.2. Figure 5.2 is an example of a plot in which only 40% of value pairs fall within this criterion, obviously an unsatisfactory state of affairs. Under these circumstances, a laboratory audit should be conducted, and substandard practices should be identified and dealt with. Re-assays can be conducted where deviation is due to analytical process error.

THE USE OF CERTIFIED REFERENCE MATERIAL AND BLANK SAMPLES

Certified reference materials (CRMs) must be obtained from accredited dealers, and due diligence must be exercised to ensure that blanks do not contain materials of interest or minerals which may interfere with the assay process. In order to minimise

HARD PLOT OF ASSAY DUPLICATES

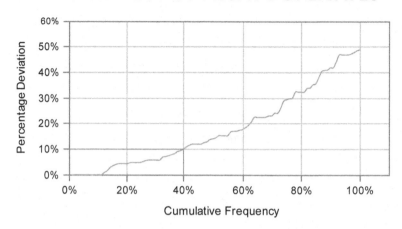

FIGURE 5.2   Example of hard plot of assay duplicates.

identification at labs, it is suggested that reference material comprises that from a similar orebody type, matrix, and colour, and that the assay technique used to determine the certified metal content is the same as the one that the assay laboratory uses.

a. Number of different grade categories

A minimum of two different metal grade categories is required, but the preferred number required are four grade categories of certified reference material. This can be accomplished, in most cases, as a direct swap of reference material with other mines.

b. Frequency of submission

At least one CRM and a blank sample must be submitted per 100 samples in the case of underground chip samples with the proviso that each panel sampled should contain a CRM and one per 20 samples in the case of drill hole samples.

NB: Under no circumstances must the CRMs be assayed separately from the samples submitted, as this is counter to the reason for submitting reference material in the first place. CRMs must follow the samples submitted in sequence to ensure that if a CRM fails, it is clear which tray is under suspicion and hence that tray can then be re-assayed and not the entire batch of samples submitted.

c. Methodology of submission

Reference material must not under any circumstances be submitted in original packaging, the reference material must be emptied into a clean sample bag with a sample ticket inserted and securely tied to prevent loss of sample and cross-contamination. Additionally, due care must be taken in sample tickets used. A sequence of tickets must not be set aside for reference material so that the reference can be easily identifiable. It is preferable that an individual person's ticket be chosen at random for submission of reference materials. As repeatability is a concern, pulp duplicates submitted must be no older than a month, as it is the precision of the laboratory at the current stage that we are attempting to identify and not a comparison to the laboratory of a year or two years ago. Under no circumstances should pulps be submitted from alternative laboratories for precision testing as this does not measure the precision of the current lab. However, pulps can be submitted to an alternative laboratory if a check on failed CRMs is required.

d. Checking of the standard deviations supplied with the reference material

Due care must be exercised to ensure that the standard deviation supplied with the reference material is the 'true standard deviation' of the assays from the mean of all samples submitted to all labs. This can be verified by obtaining the laboratory results that were used for certification and calculating the standard deviation from these. Under no circumstances must the standard deviation be calculated after the outliers have been removed. This will result in the limits for failure being of an unduly strict nature. For example, if a 90% confidence limit was used to remove outliers and then we apply a 90% confidence limit using the standard deviation obtained from the dataset with the outliers removed, one is applying a 71% confidence limit to the CRM.

e. Laboratory submissions

It is important to note that any reference material submitted does not take cognisance of the laboratories own QA/QC procedures.

f. Limits of failure

CRMs failure limits must be set at a 99% confidence limit, which is 2.326 standard deviations above and below the certified mean. In the case of blanks, this limit must be set at 4X the lowest detection limit. Please note that different analytical techniques and equipment will have different lower detection limits (which may also vary per element). Therefore, be aware of what these lower (and upper) detection limits are.

g. What to do in case of failure of assay results

In the case of failure, the entire tray on which the CRMs resides must be resubmitted for assay. If the subsequent assay fails again, the samples on that tray must be discarded, or if sufficient sample remains, this can be submitted to an alternative lab. If the CRM fails at the alternative laboratory, one needs to decide on an individual basis as to whether the CRM itself is suspect and the assay results need to be discarded or can be accepted. This can be facilitated by examining the results of all three assays, if the assays from the original laboratory and the check laboratory show a high degree of similarity, one may accept the average of the laboratory results and consider the CRM itself to be suspect.

h. Use of the historical assay data

Assay results of reference material must be graphed to determine relative accuracy of results, as well as observation of when to change reference materials if values of references have been identified by laboratories.

EXERCISING DUE CAUTION IN DETERMINING A SUSPECT ASSAY

Samples taken are not in isolation, as generally speaking, there are four to five samples per sample section. In addition, there may be four to five sample sections per area sampled. This will result in a large number of samples taken. For illustrative purposes we will assume that 25 samples were taken in the working area. In general, 25 samples will give us enough data to identify possible 'outlier' values that can be queried and re-assayed. What we must take into consideration, is that some deposits, specifically gold deposits, exhibit the nugget effect. Nuggety samples are likely to produce under estimation in terms of assays, owing to the lower occurrence of gold particles within the matrix, as each gold particle represents a larger proportion of the total gold within the sample.

If you consider that the assay laboratory will use aliquots ranging from 25 g to 50 g derived from a sample of 300 to 500 g (depending on the assay laboratory and assay methodology used), a single nugget will constitute the majority, if not the entire gold content of the whole sample. In terms of statistics, this implies that the probability of underestimating the assay in nuggety samples is around 80–90% (as this is the number of times a nugget does not enter the aliquot), whereas the probability of overestimating the assay is around 10–20%. As previously mentioned, these overestimations of the assay results are easily identified, whereas a low value returned from the laboratory is less likely to be questioned. Current thinking regarding the probability of underestimation is that a larger aliquot will reduce this problem.

OUTLIER DETERMINATION

The presence of outliers and erroneous data can have detrimental effects on summary statistics and the development of models. Therefore, it is essential to investigate any data points that appear unusual. If necessary, these points should be removed from the database in the case of erroneous data or treated appropriately in the case of outlier characteristics. In geostatistics, determining what qualifies as an outlier and when it should be treated poses challenges due to the lack of a clear logic and reasoning framework. In many cases, arbitrary thresholds or rules of thumb are commonly used to identify outliers.

Although outlier checking may appear to be a straightforward analysis, many individuals struggle to grasp the concept of what defines an outlier. At best, the understanding of outliers is vague, and at worst, it leads to a complete misrepresentation of the concept and the associated processes. Outliers refer to values or sets of values that lie completely outside the expected distribution. The occurrence of an outlier can be attributed to measurement variability or experimental errors. However, this definition poses some inherent problems as it assumes certain conditions:

- It is assumed that we possess knowledge about the characteristics of the entire distribution. This implies two conditions:
  i. We have sampled the entire population, meaning we have collected data from every individual or element in the population.

ii. During the sampling process, we have carefully and systematically divided the population into distinct subgroups.

- The term 'outlier' refers to a data point that is not substantially included in the population. In other words, it deviates significantly from the majority of the data points and does not represent the typical characteristics of the population.

- Additionally, we have a clear understanding of the correct subdivisions or categories that enable us to define individual populations in a coherent manner. This implies that we possess accurate information about how to group and classify the data based on meaningful and relevant criteria.

It is evident from the points (a) to (c) that in most cases, and particularly in the mining industry, we do not have access to the complete information required to determine outliers. Without knowing the actual distributional characteristics of the data *a priori*, it becomes challenging to identify outliers definitively. However, this does not imply that we should disregard the possibility of outliers. Instead, we need to employ probabilistic modelling rather than relying on deterministic approaches. One way to assess the potential presence of outliers is by examining the probability plot, looking for inflection points or areas where the data deviates from the expected probability line. At this stage, outliers can be determined using the distribution of the values. Typically, the upper confidence level of 99.9% is employed, which is calculated as the mean $\pm$ 3.0902 standard deviations. By plotting the data and observing whether the last interval exceeds this cut-off, we can determine if outliers exist in a particular spatial area.

If outliers are found to be clustered in a specific region, it is necessary to calculate the local mean and standard deviation within that cluster. This helps determine whether the data should be considered part of another or adjacent domain, or if it represents a genuine cluster of higher or lower values. Another approach to detect heteroscedasticity involves ranking the data from smallest to largest values. The 'F' statistic is then calculated progressively and compared to the F-test at a 95% significance level to identify any stages at which the dataset fails the test. Subsequently, the spatial examination of this data helps determine whether it belongs to a unique area or should be assigned to an adjacent domain. Emphasising the importance of checking the homoscedastic attributes of a domain is crucial, as the concept of constant variance is embedded in the logic of the kriging estimation process through the utilisation of the semi-variogram. Ensuring a consistent and valid variance assumption is fundamental for reliable spatial estimation and modelling.

Once outliers have been identified, it is crucial to determine how to utilise this information effectively. In the case where outliers form a cluster within an otherwise homogeneous domain but result in an erratic experimental semi-variogram, one possible approach is to cap the outlier values at the confidence level at which they failed. By setting an upper limit based on the confidence level, the impact of extreme outliers can be mitigated. If the capped values still pose issues with the experimental semi-variogram at this level, an alternative option is to remove them from the dataset when calculating the experimental semi-variogram.

However, it is important to note that for accurate estimation of values at their respective spatial positions, the capped values should be reintroduced into the dataset before performing kriging or any spatial interpolation method. It is essential to avoid inserting non-calculated values into the dataset, as this can lead to misleading results. The choice of the confidence level for capping or removing outliers can be subject to discussion and recalculated by an auditor. It is important to justify the selected percentage confidence level based on statistical reasoning and ensure that it aligns with accepted practices. However, arbitrary, or unsubstantiated choices should be avoided, as they can be considered fraudulent practices and compromise the integrity of the analysis.

DATA DESCRIPTION

The data description process serves several purposes in geostatistical analysis. It involves examining the dataset to assess various aspects:

a. Extent of Kurtosis: Kurtosis measures the shape of the distribution and indicates whether it has heavy tails or a peaked shape. By assessing the kurtosis, we gain insights into the data's departure from a normal distribution.

b. Extent of Skewness: Skewness measures the asymmetry of the distribution. It helps us understand whether the data is skewed to the left or right, indicating a departure from a symmetric distribution.

c. Possibility of outliers: Data description allows for the identification of potential outliers, which are values that significantly deviate from the majority of the data points. These outliers may need to be further investigated or treated appropriately, as discussed earlier.

d. Comparisons to previous modelling: Data description facilitates comparisons with previous modelling efforts. By examining summary statistics, distributional characteristics, or other relevant metrics, we can assess if there are significant differences or similarities between the current dataset and previous models. This helps to understand any changes or consistencies over time.

e. Reconciliations with previous sampling results: If there are differences in the number of samples or variations in means, variances, or other statistical measures compared to previous sampling efforts, data description allows for reconciliation and understanding of these discrepancies. It helps identify any potential reasons behind the differences and ensures the coherence and consistency of the overall analysis.

f. Final check for incorrect spatial positioning: The posting plot is an important tool for assessing the correct spatial positioning of the data. It visually displays the data points within the study area and can help identify any errors in spatial coordinates or inconsistencies in the spatial distribution. This step ensures that the data accurately represent the spatial reality and provides confidence in subsequent geostatistical analyses.

In addition to these aspects, the data description stage is where we calculate any third parameter required for log-normal distribution. This is typically done as follows: Consider the skewness of the natural logarithm-transformed data with various third parameters, denoted as Au(norm) = ln(Au[g/t] + β). The objective is to minimise the absolute value of the skewness. This transformation ensures that the log-grade values are as close to a skewness of zero as possible. By achieving a skewness of zero, we aim to make the log-grade distribution symmetric. It is worth noting that this transformation also has an impact on the kurtosis of the data. However, the objective is not to eliminate kurtosis completely, as kurtosis primarily relates to the shape of the distribution's peak in comparison to a standardised normal distribution. The focus is primarily on minimising skewness, while allowing for the presence of some kurtosis. By considering the skewness and applying appropriate transformations to the data, specifically in the context of log-normal distribution functions, we aim to achieve a more symmetric log-grade distribution. This process helps ensure that the statistical assumptions underlying the log-normal distribution are met and facilitates more accurate modelling and analysis.

HISTOGRAMS AND PROBABILITY PLOTS

In the data description phase, it is important to assess the possibility of bimodal distributions and evaluate the effectiveness of the calculated third parameters. This assessment can be done using visual tools, as shown in Figure 5.3. While analysing the data and examining the distributional characteristics, we can identify potential bimodality, which suggests the presence of two distinct modes or peaks in the data. Bimodal distributions can have significant implications for subsequent modelling and analysis, as they may indicate the existence of multiple underlying processes or sub-populations. Furthermore, probability plots are valuable in identifying outliers that may not have been previously detected in the analysis. By plotting the data against a theoretical probability distribution, such as a normal distribution, we can visually inspect any deviations or extreme values. Outliers, which are data points that fall significantly outside the expected distribution, can be identified through these probability plots. Identifying outliers is essential, as they can potentially influence the statistical analysis and modelling outcomes. Outliers may arise due to measurement

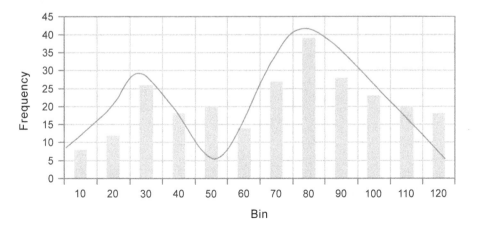

FIGURE 5.3 Histogram showing the bimodal distribution of ore grade.

errors, data entry mistakes, or genuine extreme values. By detecting outliers during the data description phase, appropriate actions can be taken, such as further investigation, treatment, or exclusion from subsequent analyses.

Upon observing the histogram example provided in Figure 5.3, where distinct indications of bimodality are present, it is crucial to consider the possibility of a mixed distribution within a single domain. This suggests the presence of two or more underlying processes or subpopulations that may be spatially distinct.

To address this, spatial identification of the mixed distribution is necessary before proceeding with further analysis. Spatial analysis techniques can help identify areas or regions within the domain where the distinct modes or populations are spatially clustered. This spatial identification is important for accurate modelling and subsequent analysis, as different processes or subpopulations may require separate treatment or modelling approaches. In addition to the histogram, probability plots, as shown in Figure 5.4, can also reveal the same distributional characteristics, such as bimodality. Probability plots provide a visual representation of the data against a theoretical probability distribution. By examining the pattern of data points relative to the expected distribution, it becomes possible to identify departures from normality or other assumptions. While utilising both the histogram and probability plot, geostatisticians can gain insights into the presence of bimodality and the need to consider mixed distributions within a single domain. Spatial identification and rectification of such mixed distributions are essential steps to ensure accurate analysis and modelling in geostatistics.

The efficacy of the third parameter value in the log-normal distribution can be assessed by comparing the normal probability plots of the data before and after the third parameter is applied. This comparison helps evaluate the impact of the third parameter on the data's conformity to a normal distribution. To illustrate this, consider Figure 5.5, which presents two graphs side by side: one without the third parameter (5.5A) and the other with the third parameter (5.5B). The normal probability plot for each case is shown. In the graph without the third parameter, 5.5A, the data points

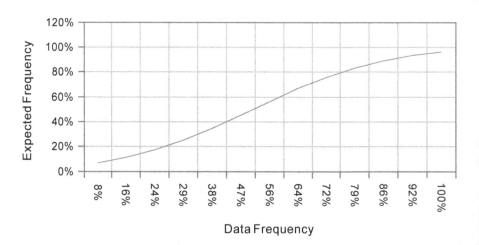

FIGURE 5.4 Example of a probability plot.

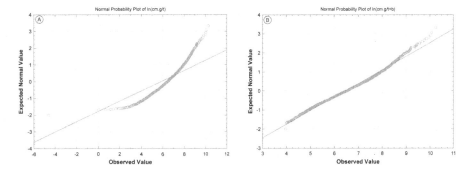

FIGURE 5.5   Example of a Q–Q plot used in understanding cross-correlated variables and in 'validating' output models. (A) has excludes third parameter, whereas (B) includes third parameter.

are plotted against the expected values from a standard normal distribution. Any deviations from a straight line indicate departures from normality. The objective is to achieve a linear pattern, indicating a good fit to the normal distribution.

On the other hand, in the graph with the third parameter applied, 5.5B, the data points are plotted after the transformation or adjustment using the calculated third parameter. This adjustment aims to minimise skewness and make the log-grade distribution as close to a skewness of zero as possible, as discussed earlier. By examining this transformed plot, we can assess whether the inclusion of the third parameter has improved the data's conformity to a normal distribution. Comparing the two graphs visually allows us to determine whether the inclusion of the third parameter has enhanced the normality of the data. A more linear pattern in the graph with the third parameter, 5.5B, indicates a better fit to the normal distribution, suggesting the efficacy of the chosen third parameter value.

As can be seen, the addition of the third parameter improves the look of the Normal probability plot showing quiet clearly that the distribution in question follows a three-parameter log-normal distribution.

CHECK FOR TRENDS

In geostatistical analysis, the concept of trend refers to a consistent change in the value of a variable as it varies across space. Identifying any underlying trend in the data is crucial at an early stage of the analysis process, as it can significantly influence subsequent steps such as domaining and modelling. One common approach to check for trend is by examining the residuals and the goodness of fit of polynomial trend surfaces. Trend surfaces are mathematical functions that capture the spatial variation of the variable under consideration. These surfaces can be expressed as polynomial equations of different orders, such as linear, quadratic, or cubic trends. In practice, it is accepted that if no significant trend is observed on linear, quadratic, or cubic trend surfaces, the data is assumed to be trendless. This means that the variation in the data across space does not exhibit a consistent and systematic pattern that can be represented by a polynomial trend surface. To assess the presence of trend, the residuals, which are the differences between the observed values and the predicted values from the trend surface, are examined. If the residuals do not display any discernible pattern or exhibit significant departure from randomness, it suggests the absence of trend in the data.

Furthermore, the goodness of fit measures, such as the coefficient of determination (R-squared) or the analysis of variance (ANOVA), can provide additional insights into the adequacy of the polynomial trend surface in capturing the variation in the data. While it is theoretically possible to check for trend using polynomials of any order, it is common practice to assess trend up to cubic order. If no significant trend is observed on these lower-order trend surfaces, it is generally assumed that the data is trendless. Identifying and understanding the trend in the data early in the analysis process is essential for accurate modelling and domain identification. It ensures that subsequent geostatistical techniques are appropriately applied to account for the underlying spatial patterns and variations in the data.

    a. Significance testing or what are the chances of?

       At this point, it is prudent to consider the process of significance. Given enough criteria, it is possible to sub-divide any data set down to individual sample level and lower. Here we are faced with the question of whether the subdivision one contemplating will have an undesirable outcome on the final estimation process. As such, we need to consider the estimation process envisaged and the constraining criteria of such process. The fact that variables may have different values does not constitute a basis for subdivision; it is for this reason that estimation techniques were developed in the first place; that is, to estimate the occurrence of differences in values; if all the values were the same, the estimation process would be unnecessary. On the other hand, most estimation processes work on the assumptions that:

    b. The samples within a domain consist of a single distribution (unimodal).

    c. The variance within a domain remains constant (homoscedastic).

Consider the example in relation to point c in Table 5.1.

In this case, the means differ by over 2,000, the variances are the same. Additionally, a test of means (e.g. t-test[2]) indicates that at a 99.9% confidence level, that the two sets of samples can be from the same distribution. That is not to say that one cannot infer that the two sample sets are from different distributions, it merely means doing so has a 99.9% chance of being wrong.

It is with this in mind that one considers the case of trend and its significance. Most geostatistical software packages include a method for polynomial trend surface fitting; however, not all packages include the necessary statistics to determine whether the trend is statistically significant. Nevertheless, most packages will provide some sort of summary of the ANOVA, as shown in Table 5.2. Note that the t-test is a statistical hypothesis test used to determine if there is a significant difference between the means of two groups or populations. It is particularly useful when working with small sample sizes and when the population standard deviation is unknown. The t-test calculates a t-value based on the means, standard deviations, and sample sizes of the two groups, and compares it to a critical value to determine statistical significance.

TABLE 5.1
Example of homoscedastic distribution

Feature	Example A	Example B
Sample 1	425	2,650
Sample 2	850	3,075
Sample 3	1,250	3,475
Mean	842	3,067
Variance	170,208	170,208

TABLE 5.2
Example of analysis of variance (ANOVA)

Source of Variation	Sum of Squares	Degrees of Freedom	Mean Squares	F-Test
Regression	2,558,554	9	284,283	
				0.9703
Deviation	20,508,870	70	292,983	
Total variance	23,067,424	79		
Goodness of fit	0.1109			
Correlation coefficient	0.3330			

From Table 5.2, it can be observed that the correlation coefficient is 0.333, which indicates a moderate correlation. While it is not considered high, it still suggests a relationship between the variables being analysed. Additionally, it is evident that approximately 11% of the total variation of the dataset can be explained by the trend alone. These two statistics provide insights into the contribution of the trend to the overall variation. However, determining the significance of the observed trend requires further analysis. One commonly used statistical test is the F-test. Through consulting the F distribution tables, we can evaluate the significance of the trend at a certain confidence level. In this case, at a 90% confidence level, the critical F statistic value is 0.5805. Comparing this value to the computed F statistic of 0.9703 obtained from the data, we find that the trend observed is not significant with a 90% confidence level.

In cases where the table mentioned is not provided, it is advisable to subdivide the dataset based on the areas of trend or other relevant factors. After subdividing, significance testing of the variances can be performed within each of the domains to assess the significance of the trends observed in those specific subsets.

DOMAINS

To accurately estimate the metal grade or quality of a mineral deposit, a common practice in mineral resource estimation is the partitioning of the deposit into spatially separable domains, a process known as 'domaining' (Larrondo and Deutsch, 2005; Yunsel and Ersoy, 2011). This approach incorporates geological control on the distribution of grades by creating distinct domains within the deposit. When establishing these domains, several parameters must be considered, including lithology, structure, alteration, and grade. The information pertaining to these parameters is utilised to assign each sample to a unique domain. A domain identifier, typically a number, is then assigned to the sample, enabling the software used for analysis to identify the sample and its associated domain. To ensure the validity of the domains, statistical tests such as F-testing and coefficient of variation tests are applied. When domains exhibit similar characteristics, merging them is advisable to optimise the representation of the deposit. It is important to minimise the number of domains used while still maintaining an accurate representation of the deposit.

As more information becomes available, the domains must be continually reassessed and refined. This iterative process allows for the integration of new data and ensures that the domains remain up-to-date and aligned with the evolving understanding of the deposit. Geological domains serve a meaningful purpose only if they contribute to the accurate estimation of the mineral of interest. Consequently, these domains must adhere to the constraint of providing quasi-stationary domains.

> In geostatistics, quasi-stationarity refers to the condition where a random variable exhibits spatial dependence that is approximately stationary over a certain range of distances.

Stationarity is a fundamental assumption in geostatistics, positing that the statistical properties of a variable, such as mean and variance, do not vary with location within a study area. While complete stationarity is often not observed in practice, quasi-stationarity allows for the application of various geostatistical techniques that assume a level of stationarity.

Through defining geological domains that exhibit quasi-stationary behaviour, geostatistical techniques such as variograms or kriging can be effectively employed. The assumption of quasi-stationarity implies that the variogram model or spatial correlation structure remains reasonably consistent within localised regions or neighbourhoods of the study area. This local approximation of stationarity facilitates the estimation and prediction of values at unsampled locations based on nearby observations. Given the complexity and interdisciplinary nature of establishing geological domains, it is crucial to involve both the geology team and the mineral resource estimator in the process. Their collaborative effort ensures that the domains accurately reflect the geological characteristics of the deposit while meeting the requirements for mineral resource estimation.

ESTIMATION OF STATISTICAL DOMAINS

If the geological domains have been adequately defined, the statistical domains should only exhibit statistical differences, such as variations in sub-distributions within each individual domain. To assess these differences, the F-test should be performed on

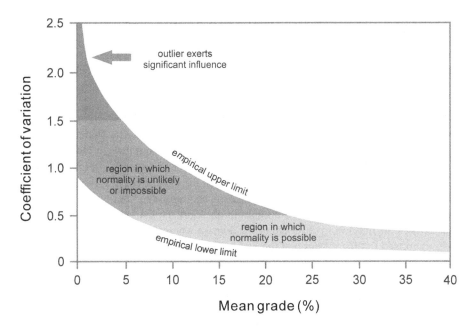

FIGURE 5.6 Coefficient of variation intervals versus percentage mean grade representing variability decay.

samples within each geological domain, and the resulting test statistics should be analysed for second-moment stationarity. To conduct this analysis, we first assume a uniform distribution and evaluate each sample's similarity of variance to the initial group or distribution using the *F*-value. It is anticipated that the variance of a sub-section of an orebody will likely be smaller than that of the entire orebody (Krige, 1951, 1997). The determination of stationarity is context-dependent and subject to interpretation, thus lacking consensus on the precise conditions and confidence levels for assigning domain clusters (Dias and Deutsch, 2022).

Typically, a 95% confidence limit ($\alpha = 0.05$) is employed, where each sampled point is assigned to a domain based on the statistical significance of variance until all points have been allocated to a specific unique domain. This assignment is based on the hypothesis $Ha = \sigma^2_1 \neq \sigma^2_2$, where Ho represents no difference in variances and Ha denotes a difference in variances. For each domain, the coefficient of variation (COV or CV) was employed to quantify the dispersion of data points around the mean within that domain (Isaaks and Srivastava, 1989). Thus, Equation 5.1 was utilised to produce Figure 5.6:

$$CV = \frac{\text{Standard Deviation}(\sigma)}{\text{Domain Mean}(\mu)} \tag{5.1}$$

DATA DECLUSTERING AND SUPPORT SIZE DETERMINATION

An estimate of the mean of a particular area of interest is frequently required during the estimation process. This, in turn, needs to be as representative as possible of the total area in question and not just a portion thereof. As a result of collecting data from areas where values are high or low, data is often biased and not representative of the volume in those areas. To this end, one needs to consider a method of removing the effect of clusters of data that frequently occur. Consider Figure 5.7 which is a worst-case scenario example of the bias introduced due to the clustering of samples.

As can be observed, the area in the top left-hand corner of the block has 63% of the sampling, whereas it represents only 5% of the area spatially. Under these circumstances, there is a definite chance that any mean used from the sampling data has a distinct chance of being incorrect. Thus, we need to ensure that the weight assigned to the clustered area is representative of the area the cluster covers. To achieve this objective, one needs to consider the relationship between the area of clustering and the remaining area of the zone or domain in question. This needs to be a block size that ensures the maximum coverage of the domain where data exists; due care needs to be exercised in this respect that the less well-informed areas get an equal weighting throughout the area of interest. Consider the graphic in Figure 5.8.

In Figure 5.8A, one observes pronounced clustering in the top left-hand corner with the result that this area receives the majority on the weight whereas the remaining area receives the least. In Figure 5.8B, the top left-hand cluster is evident in five of the blocks and as such still carries over 45% of the weight for the mean, whereas the actual area of representativity is far less than this. In Figure 5.8C, the cluster is found

FIGURE 5.7   Example of clustered samples.

in four blocks, an improvement but still significantly overrepresented. Figure 5.8D indicates the idealised situation, with the cluster weight being equal to its area of representativity as well as having most blocks represented with data. In Figure 5.8E, one observes that the top left-hand corner is now once again receiving an increasing amount of weight and therefore is being overrepresented in the mean. Finally, Figure 5.8F indicates a block the size of the area of interest with the result that the mean becomes that indicated in Figure 5.8A once again. From the above, it is easy to deduce that in any declustering exercise, one will start with an un-clustered mean and end with an un-clustered mean. In between the two scenarios, the mean will deviate from the un-clustered mean increasingly as one approaches the optimal declustered configuration and will converge once again on the un-clustered mean as soon as this position is passed. This, in turn, mathematically translates into a quadratic form, as shown in Figure 5.9, with the optimal declustered position sitting at the turning point of the graph. This is easily obtained by taking the first derivative of the least squares fit of the graph to the declustered figures. Some may contend that the declustered mean is now represented by a mixed support mean. This, however, can be countered by pointing out that if each block had only one sample, and that sample was randomly chosen from any position within the block, one has a chance that the value obtained may very well closely emulate the average value of that block. Because each average of the blocks at the optimal block size is treated as a single point within the block, the likelihood of obtaining a single sample that approaches the value of the block's mean is high. Any potential bias due to differing support sizes can be considered as being negligible. However, the same cannot be the case when each individual block has its own weight.

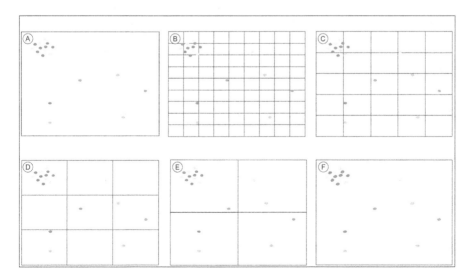

FIGURE 5.8 Example of clustering of samples that frequently occurs within the mining environment (known as preferential sampling), with areas of subjective interest often being oversampled to 'prove up or down'.

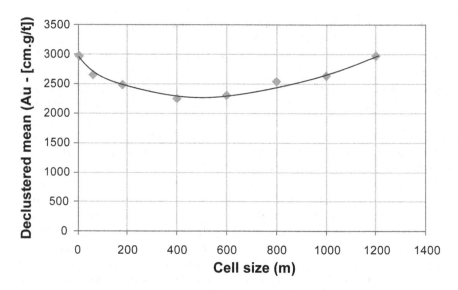

FIGURE 5.9   Example of regression for the best-declustered mean.

DERIVING THE ACTUAL SELECTIVE MINING UNIT (SMU)

Immaterial of what anyone may perceive as being the SMU, the actual SMU is represented by the configuration remaining after all mining has been completed as shown in Figure 5.10. This represents the actual level of selectivity obtained by the mining process and is totally independent of the apparent level of selectivity monthly. Here, it is possible to get the impression that there is a higher (or lower, in some cases) level of selectivity than exists once the entire area of interest has been fully exploited. Here are a few examples to help you understand the concepts at hand (Figure 5.10).

When considering the above example, it is obvious that the means of access to the unmined area is via one of the two raises and as such the mining configuration for the first 10 metres for both the raises is entirely possible. However, because the final configuration for a 10 m × 10 m SMU must look like the sketch in Figure 5.10, it begs the question of how one moves from one white block to another while the blue blocks remain solid ground. Examining the sketch in Figure 5.11, one observes that for two raises 80 m apart at best the SMU obtainable is a 40 × 10 m.

From the sketch above, one can observe that with a 40 m × 10 m SMU, all configurations of mining (although they may not be easy) are possible, whereas with the 10 by 10 m SMU, actual mining of any block, with raised space of more than 20 m apart is *practically impossible*. From the previous two simple yet effective examples, it should be clear that one of the dimensions of the SMU is defined by half of the distance between access points. The other dimension of the SMU could be one of two criteria:

a. If the deposit is extremely shallow and no rock engineering constraints apply, this could be of the minimum size technically feasible to mine a panel or portion of ground (in other words, the panel length).
b. If rock engineering constraints apply, the minimum size of the pillar can be left in situ while the surrounding area is mined out.

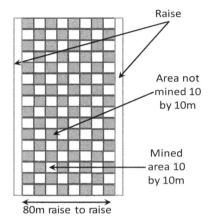

FIGURE 5.10   Example of the 10 × 10 m SMU fallacy.

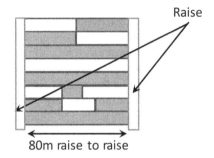

FIGURE 5.11   Example of the 40 × 10 m SMU fallacy.

In the vertical dimension, the same arguments will apply if the SMU is less than the thickness of the package, then the assumption is being made that some of the ground between cuts is going to be left *in situ*.

Three different criteria need to be taken into consideration to determine whether an SMU is practically feasible when it comes to the actual mining scenario:

    a. What are the technical constraints on mining to this SMU?
    b. What are the rock engineering constraints?
    c. As in the case shown in the first sketch under this section. Is it even possible to gain access from one area to another if the suggested SMU were to be adopted?

A perfect example of the ridiculous is the SMU that is quoted on a mine as being 10 m × 10 m × 1 m. As seen in Figure 5.10, it is impossible to mine at 10 × 10 m grid when the raise spacing is more than 20 m apart; however, to say further that this mine is going to mine a single metre in the vertical and then leave a metre and then once again mine 1 m, etc., shows an absolute lack of understanding of the selectivity process and mining in general. In any event, under no circumstances should one mix up the concepts of resolution in estimation and the SMU. The former is the measure of how accurately one can demarcate economic areas from uneconomic areas, the concept of the SMU is the indication of how effectively one can safely and productively mine the economic areas only, while leaving the uneconomic areas behind without hindering the process of mining other economic areas that may be beyond the current working areas.

INFORMATION EFFECT

Simply stated, the information effect, is the error made in the estimation of the grade estimation of an SMU-sized block at the time of decision making. To clarify the phrase, we will expand slightly on the term 'at the time of decision making'. Here we mean the time at which we make the final decision as to mine or not to mine and then immediately thereafter commence mining. As can be seen, this is different from the error in estimation of a block of ground that is to be mined ten years in the future. One may ask how it is that we know what error we will make on a block when the decision to mine or not may only occur a few years into the future. We do not know, but we can estimate it by using the planned expected sampling or drilling configuration at the time of decision making and the semi-variogram modelled for the area. The process is quite simple and entails using the planned sampling or drilling configuration as a template and then kriging an SMU-sized block using the relevant domains semi-variogram. The resultant kriging variance is the estimation variance required for the information effect. This can be accomplished by setting up a data template and using the modelled semi-variograms and a single block model prototype in a package such as Datamine Studio™ to kriging a single block and obtain its corresponding kriging variance.

Note: If you have transformed the sampling data into log space, the corresponding estimation variance must also be in log space.

Whatever your methodology, it should be well documented, and the blocks estimated per domain to determine the information effect and the corresponding semi-variogram models used should be available for scrutiny.

CALCULATION OF EXPERIMENTAL SEMI-VARIOGRAM

A semi-variogram is a mathematical tool used in geostatistics to assess the spatial variability or dissimilarity of a phenomenon across a given area. It provides valuable insights into the spatial correlation structure of data points in a spatial domain. The semi-variogram is calculated by measuring the squared differences in values between

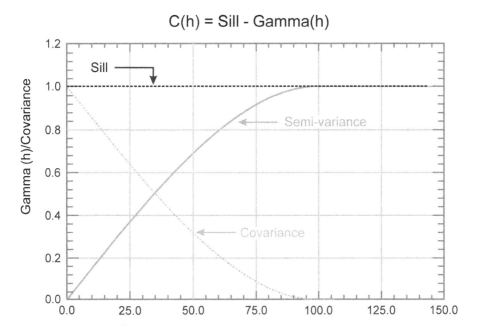

FIGURE 5.12   Illustration of Equation 5.3.

pairs of data points at varying distances and directions. To compute a semi-variogram, we start by selecting a lag distance, which determines the maximum separation between data points that are considered for comparison (Equation 5.2). Next, we calculate the differences in the values of data points separated by this lag distance vector (comprising both distance and direction). These differences are then squared, representing the squared dissimilarity between the data points. The semi-variogram is constructed by averaging these squared differences across all pairs of data points separated by the lag distance, h, quantified using a semi-variogram (Krige, 1997), thus (Figure 5.12):

$$\gamma(h) = \frac{1}{2N(h)} \sum_{i=1}^{N(h)} \left(Z(u_i) - Z(u_i + h)\right)^2 \tag{5.2}$$

where $\gamma(h)$ is a measure of dissimilarity (or spatial variance) versus distance, $N(h)$ corresponds to the number of all data point pairs separated by h, u_i corresponds to a data point in 2D or 3D space at location i, and $Z(u_i)$ is the value at point u_i. The $1/2N(h)$ is used so that the covariance function ($C(h)$) and variogram may be related directly:

Thus,

$$C(h) = \sigma^2 - \gamma(h) \tag{5.3}$$

where σ^2 corresponds to the total sill (C) or total variance of the samples.

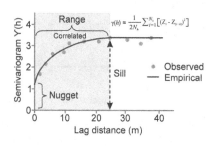
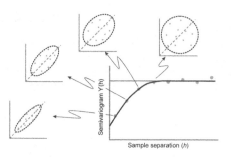

FIGURE 5.13 Example of an experimental semi-variogram fitted with a spherical model.

Nugget effect: It is the typical difference between the samples where we must take samples almost adjacent to each other. Imagine splitting core and submitting both halves. The difference between the grade of the core halves would be the nugget effect. Precious metals are expected to have a higher nugget effect than some of the based metals (Figure 5.13).

Sill: As the separation between samples increases, so the difference between them increases, until we reach a distance beyond which the difference between sample grades is not dependent on their separation but is the same as the background variability or the population variance. This plateau in the variogram value is called the sill.

Range: The range is the distance beyond which the samples are no longer spatially correlated. The range is the distance at which the total sill is reached.

When using the experimental semi-variogram it is of utmost importance that one should understand the pros and cons of several types of semi-variograms. Here, we will consider the relative benefits of the point and block experimental semi-variograms.

THE POINT SEMI-VARIOGRAM

We must take the following into consideration.

a. Data density. Here, the perceived data density corresponds directly to the actual data density.
b. Values in space. Here, the value indicated at a certain position in space (assuming the position was correctly captured, and the value correctly obtained) corresponds to the actual value at that position.
c. Will be the shortest range of influence of any of the semi-variograms; therefore, has limited distance at which estimation may occur from data.
d. Will have the greatest overall variance of all semi-variograms.
e. Can be the most erratic of all semi-variogram structures.

THE BLOCK SEMI-VARIOGRAM

Here we must take the following into consideration.

f. Data density. Here the perceived data density does not correspond directly to the actual data density since some of the blocks may not be totally covered by sampling.

g. Values in space. Here the value indicated at a certain position in space (assuming the position was correctly captured, and the value correctly obtained) may not correspond to the actual value at that position due to the centre of gravity of samples or the centre of block being used as the coordinates of the samples. Preferable to utilise coordinate centre of gravity as this at least attaches some of the sample's attributes to the spatial position of the block.

h. Will be a longer range of influence than the point semi-variograms. Theoretically, the total sill increase should equal to the range of the point semi-variogram plus the size of the block. However, very often the range of influence increases by much more than this due to much of the variability being taken up inside the block and hence structure not noted originally now becomes evident. Nevertheless, as a rule, further away from data estimation becomes possible due to the increased range.

i. While point (h) does allow for estimation further away from the source data, one must still consider that each block may not be representative of its correct mean due to a limited number of samples less than that required for a correct mean within the block averages. We will deal with this aspect and how to compensate for this later in the document.

j. To avoid production of ordinary kriging estimates with negative kriging weights of all the blocks and less than that of a point semi-variogram, will have overall variance equal to the variance.

k. Will be less erratic than point semi-variograms. There is, however, a proviso which is that there are sufficient block pairs for each lag. If this is not met an extremely erratic experimental semi-variogram may result.

SEMI-VARIOGRAM MAPS

Normally the initial calculation is in the order of six primary directions with say a 30-degree increment. Subsequently, the semi-variogram contours are generated and examined to determine the direction of anisotropy. Here due cognisance must be taken of the contour that corresponds to the sill of the semi-variogram, that is the variance of the data. This is especially important when there appear to be two or more directions of anisotropy with different ranges; in these cases, the direction with the largest range that reaches the sill must be used. Consider the contour plot example (Figure 5.14).

If one was considering the previous contour plot based purely on the contours one may become confused as to what the direction of anisotropy is (the shorter line or the longer). However, as soon as one takes into consideration that the sill of the semi-variogram is 0.35, it becomes clear that the direction of anisotropy is represented by the purple line. Once the correct direction of anisotropy has been determined re–calculate the experimental semi-variograms using these directions as the primary directions for calculations.

SEMI-VARIOGRAM MODELS

The kriging and simulation algorithms require a positive-definite model of spatial variability; hence, the experimental semi-variogram cannot be used directly. Instead,

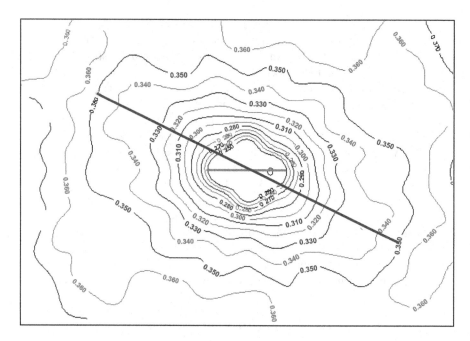

FIGURE 5.14   Example of a semi-variogram map.

a model must be fitted to the data to approximately describe the spatial continuity of the measured variables. A positive-definite semi-variogram models ensures that for all possible data and estimate configurations, there are no paradoxes. Here are the factors to be considered:

a. The number of sample pairs for each lag.
b. The type of model needed.
c. Whether or not the model needs more than a single structure.
d. The sample variance used for the semi-variogram calculation.

Considering the factors in the order in which they appear in the list above. When there are a limited number of pairs available for a certain lag, the wrong impression could be obtained as to how to model the experimental semi-variogram. This occurs most often in the case of the first lag, which may be less than the average sample spacing and thus have a limited number of pairs available; however, it may occur under other circumstances and lag spacings throughout the experimental semi-variogram. The question most asked is how many pairs is enough? The answer is not simple. A general rule could be that any lag that has less than 20% of the average number of pairs should not be taken into consideration; however, there is a proviso. There needs to be sufficient lags available to model the semi-variogram structure and in some cases strictly applying the 20% rule results in insufficient points necessary to model. This can occur with a high degree of frequency in areas of sparse data were the lags closer to the origin have less data as shown in Figure 5.15).

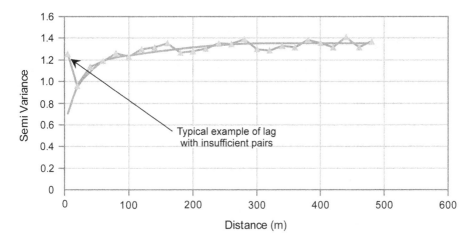

FIGURE 5.15 Example of a semi-variogram with insufficient samples.

Although in most cases, one will encounter that a spherical semi-variogram model will suffice (see Equation 5.4), it is important to know that not all structures will adhere to this type of model (Figure 5.16). Mathematically, the spherical semi-variogram model can be given by Equation 5.4:

$$\gamma(h) = \begin{cases} C_0, & h = 0 \\ C_0 + C\left(1.5\dfrac{h}{a} - 0.5\left(\dfrac{h}{a}\right)^3\right), & 0 < h \le a \\ C_0 + C(h), & h > a \end{cases} \tag{5.4}$$

where a is the effective range of influence in coordinate units, $C_0 + C$ is the sill, and C_0 is the nugget.

The spherical semi-variogram model has a nugget effect of $c_0 \ge 0$, a range parameter of $a > 0$, and a finite sill value of $C_0 + C$, where $C > 0$. This permits the calculation of the covariance function, see Equation 5.3.

Table 5.3 illustrate the different semi-variogram models. The following are some examples of common semi-variogram models one may encounter (Table 5.3 and Figures 5.16, 5.17, and 5.18).

In all circumstances, one needs to exercise caution when selecting a type of model to apply. If, for example, one is considering a mine in which all the semi-variograms up until now have been spherical, one must view with suspicion an experimental semi-variogram that exhibits an exponential structure. A particular example of this occurrence is when one expects a spherical or another type of model and finds the experimental semi-variogram exhibiting a power function. This indicates the presence of a trend and under no circumstances should one attempt to fit a power model to the data.

TABLE 5.3

Equations for most common semi-variogram models

Semi-variogram Model	Equation
Spherical	$\gamma(h)=c\left[1.5\dfrac{h}{a}-0.5\left(\dfrac{h}{a}\right)^{3}\right]$
Exponential	$\gamma(h)=c\left[1-\exp\dfrac{-3h}{a}\right]$
Gaussian	$\gamma(h)=c\left[1-\exp\left(\dfrac{2h}{a}\right)^{2}\right]$
Power Model	$\gamma(h)=ch^{\omega}$
Hole Effect	$\gamma(h)=c\left[1-\cos\left(\dfrac{h}{a}\pi\right)\right]$

Figure 5.18 demonstrate an example of this phenomenon.

Contrary to widespread belief most semi-variograms cannot be modelled using a single structure. It is important to take into consideration the underlying principle behind modelling a semi-variogram in the first place. That is, to provide a model that as closely emulates the actual data variance across a distance as closely as possible. This can only be accomplished by using first, a model that is appropriate for the data and second, one that has more than a single component if it is required. Following are examples of single and multiple structure semi-variogram models (Figure 5.18).

One can clearly see that although both examples have the same nugget effect, overall sill and range, the single structure model (additional line on graph B) is a totally inappropriate model for the data. One could argue that by reducing the range a better fit could be accomplished; however, this would simply mean that the range would have to be greatly reduced and that there would be a mismatch of data to the semi-variogram at the top of the model rather than at the bottom (Figure 5.19).

Due care must be exercised in checking that the overall sill used is obtained from the variance of the data used to calculate the semi-variogram. If the data does not reach the sill, consider recalculating the experimental semi-variogram with more lags (to a greater distance).

In the case of thick deposits, it is often the accepted practice to utilise the borehole semi-variogram to obtain the nugget effect, as the sample spacing in this direction is generally of shorter distances. Although, in principle, there is nothing wrong with

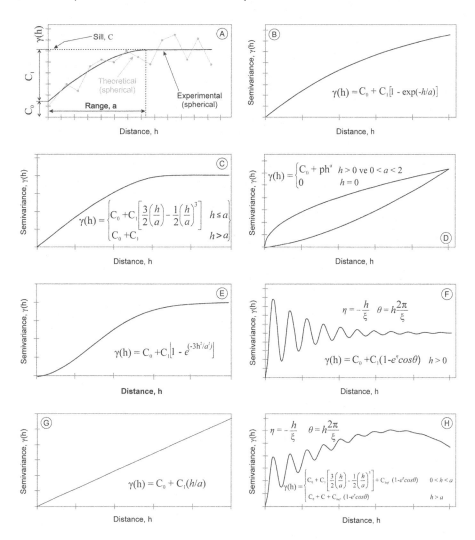

FIGURE 5.16 Example of the most common theoretical semi-variogram models.

this practice, due care must be taken in liaising with the geology department in ascertaining whether:

a. The deposition in the vertical direction is continuous.
b. One is not modelling across different reefs.
c. There are no discontinuities in the vertical direction.
d. One is not modelling a vertical trend.

Where the vertical thickness of the domain being modelled is less than twice the composite length, using the borehole semi-variogram to obtain the nugget should be

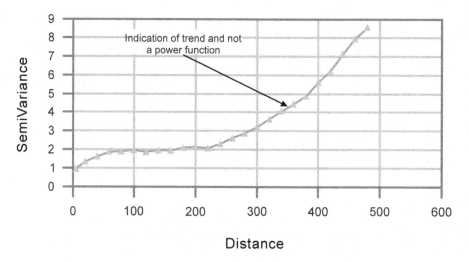

FIGURE 5.17   Example of semi-variogram showing trend.

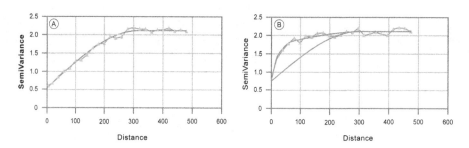

FIGURE 5.18   Example of semi-variograms with single (A) and multiple (B) structures.

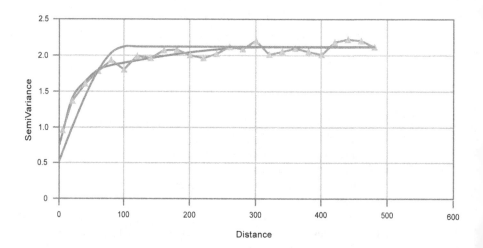

FIGURE 5.19   Example of semi-variogram showing multiple structures.

avoided as it will result in a nugget effect which will be too high because half of the pairs will be at a different support size.

KRIGING

Given a support (S) for the estimator ($\hat{Z}_S(u_0)$), where the $S \subset d$, a neighbourhood around the point (u_i) with their associated weights (w_i), the block estimator can be given as Equation 5.5:

$$\hat{Z}_S(u_0) = \sum_{i=1}^{n} w_i Z(u_i) + \text{Error} \qquad (5.5)$$

subject to $\sum_{i=1}^{n} w_i = 1$ (Olea, 1999). The weights are derived from the discretised semi-variogram model, meaning that their values are dependent only on the distances and directions between the control (sampled) points but not on their concentration values. Point semi-variograms must be discretised when used for point-to-block estimation. The kriging system is applying a volume–variance correction by taking the average of the variability between a sample and all discretisation points within a block. The discretisation points within small blocks will be closer together than discretisation points within larger blocks. This means a more similar span of separation distances between sample and discretisation point and, therefore, more similar variogram values. In contrast, discretisation points within larger blocks are further apart and thus use a wider range of variogram values to calculate the average variability between the sample point and the block.

In ordinary kriging, the local mean is not known, and the kriging variance is minimised subject to the unbiasedness constraint that the sum of the weights is equal to unity (Isaaks, 2005). A Lagrange multiplier (λ) is used to create an objective function of the pseudo-unconstrained optimisation problem (to minimise the kriging variance) (Olea, 1999).

This relaxes the assumption of global stationarity since the mean can be effectively estimated using the local search data. Kriging outputs include the kriging estimate and its variance, which is a measure of the estimation uncertainty (Deutsch et al., 2014). The kriging variance is zero where the kriging estimate occurs exactly at a data location for point estimates. Estimates further away from support points result in higher kriging variances. Interpolation usually results in unavoidable smoothing and for kriging, estimation away from supporting data can result in smoothed estimates if the estimation relies on excessive amount of data points. Here, we provide a simplified formulation. For kriging estimation of a block (X_B) with a support (S), a point-to-point covariance ($\text{Cov}(\ )()$) and a point-to-block variance:

$$Cv(x_p, S) = \frac{1}{S} \int_S \text{Cov}(x_p, x) dx \qquad (5.6)$$

a Lagrange multiplier (λ) used to minimise the kriging variance and a field of scalars ($Z_n(X_n)$), the system of equations of the weights (w_i) of kriged estimates ($Z_S(X_B) = \mathring{a}w_i Z(x_i)$) can be written as:

$$\left\{ \begin{array}{c} \displaystyle\sum_{j=1}^{n} w_j \, \mathrm{Cov}\left(x_j, x_k\right) + \lambda = Cv\left(x_k, S\right), \quad k = 1, \ldots, n \\[3mm] \displaystyle\sum_{j=1}^{n} w_j = 1 \end{array} \right\} \tag{5.7}$$

The block estimation variance can be given by Equation 5.8:

$$\sigma_B^2(X_B) = CV\left(S_i, \, S_j\right) - \sum_{j=1}^{n} w_j Cv\left(x_j, S\right) - \lambda, \tag{5.8}$$

where

$$CV\left(S_i, \, S_j\right) = \frac{1}{S_i S_j} \iint\limits_{S_i \, S_j} \mathrm{Cov}\left(x_1, x_2\right) dx_1 dx_2$$

denotes the block variance. In the instance that the kriging applies to a point instead of a block, the covariance functions are replaced by a semi-variogram model, and the support converges to a local neighbourhood. The kriging variance for a point estimate can be given by Equations 5.9:

$$\sigma_K^2(x_p) = \sum_{j=1}^{n} w_i \gamma\left(x_p, x_j\right) - \lambda \tag{5.9}$$

where $\gamma(x_1, x_2) = \gamma(h(x_1, x_2))$ is a semi-variogram model. It is important to measure the performance of kriging. Once a block is estimated, several statistics can be calculated including kriging variance, kriging efficiency (KEFF) (Krige, 1997) and slope of regression (SLOR) (Snowden, 2001). KEFF, as a measure of the efficiency of block estimates, is a ratio of kriging variance σ_K^2 (estimation variance) normalised by the variance of the true blocks σ_B^2 (Deutsch and Deutsch, 2012). The KEFF can be expressed as Equation 5.10:

$$\mathrm{KEFF} = \frac{\sigma_B^2 - \sigma_K^2}{\sigma_B^2} \tag{5.10}$$

A high KEFF means that the kriging variance is low, and the variance of the block estimates is equal to the variance of the true block values. The value of KEFF depends on sample configuration and is expressed in fractions of 0–1. The kriging variance varies from block to block, which implies that the KEFF varies similarly. There are several limiting cases discussed by Krige (1997). For perfect estimations, the KEFF is 1. However, it can be negative if the kriging variance is greater than the true block variance, where the estimation variance exceeds the block variance. Where a block has a KEFF of 1, we expect a perfect match between the estimated and true grade distributions. As data become sparser or clustered, or as blocks are extrapolated more than interpolated, KEFF drops. Sometimes KEFF can be negative, signalling extremely unreliable estimates.

The SLOR can be used to assess the performance of kriging by approximating the conditional bias of the kriging estimate. It can be written as Equation 5.11:

$$\text{SLOR} = \frac{\sigma_B{}^2 - \sigma_K{}^2 + \lambda}{\sigma_B{}^2 - \sigma_K{}^2 + 2\lambda} \qquad (5.11)$$

where λ is the Lagrange multiplier (Snowden, 2001). For simple kriging, SLOR is always exactly 1, whereas for ordinary kriging, the SLOR is less than 1. This implies that ordinary kriging is conditionally biased in general unless SLOR is 1. When the slope is 1.0, the estimated high-grades and estimated low-grades correspond accurately to the respective true high and low-grades. The magnitude of the conditional bias can be understood by making a spatial map of SLOR (Isaaks, 2005).

Krige (1997) deemed this a kriging anomaly and states that valuing the block with the mean would be more efficient, assuming a priori knowledge of the sample mean for a particular estimate. The SLOR depends on the Lagrange multiplier, block variance and kriging variance and is a dimensionless quantity that is expressed in fractions of 0–1. The regression of the true values given the estimate is an indication of the conditional bias in the estimate. For simple kriging, SLOR is exactly 1. This is achieved precisely in simple kriging by reducing the variance of the estimates that is, smoothing. In ordinary kriging, in which the Lagrange multiplier is used, SLOR is less than 1. Hence, the ordinary kriging estimates of a block are more conditionally biased (Figure 5.20), as the regression slope deviates further from 1. The block size can be increased and/ or estimated through simple kriging instead of ordinary kriging to resolve excessive conditional bias (Figure 5.20). However, estimating into larger blocks can produce an over-smoothed distribution and does not provide the resolution required to select blocks for mining (Figures 5.21).

For cross-validation, three schemes are commonly used, namely:

a. pre-modelling cross-validation to optimise kriging parameters,
b. leave-one-out or jack-knifing, which is an out-of-sample testing method that hides a subset of the data from modelling (a test set), and all calculations (estimation of the mean, variance, and variogram) are performed without the test set; then the prediction is compared with the test set (Figure 5.22),
c. swath analysis that compares two sets of data, for example, sample grades used for the estimation and the corresponding estimate. If a swath analysis

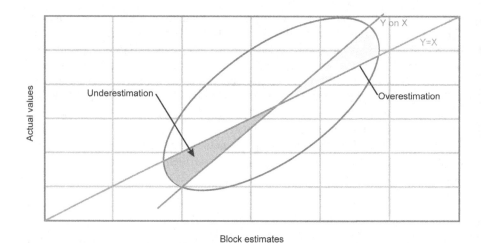

FIGURE 5.20   Example of under- and over-estimation of values due to conditional bias.

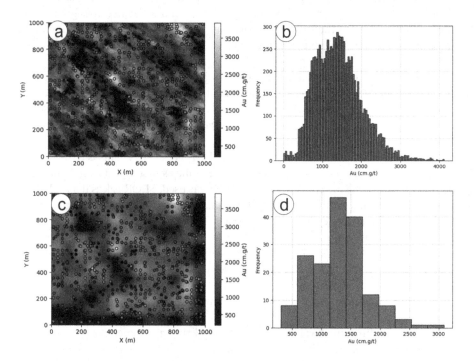

FIGURE 5.21   Example of ideal 2D block model showing ore variability versus smoothed Kriged model.

indicates that the sample data and the model's estimate match well, then it is more likely that the block estimates from the individual samples would also be correct.

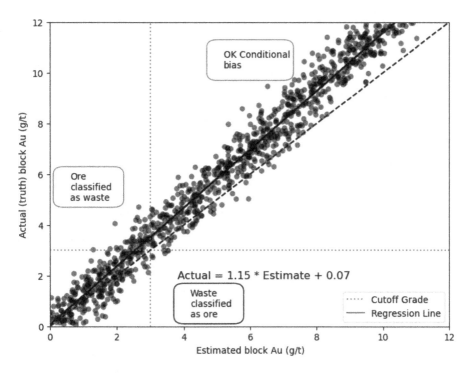

FIGURE 5.22   Example of leave-one-out cross-validation.

Jack-knifing is particularly suitable for data-dense areas, and use as much data as possible to check estimation parameters such as:

- Infill drill programmes
- Twin/wedge drilling
- Reconciliation data
- Robustness of estimation parameters

In addition to kriging, continuous various, categorical, and censored data can be estimated using indicator kriging. The process of indicator kriging is one that supplies an indicator of true or false for each piece of data for a certain variable. This could be a straightforward true or false such as a sample being within a certain facies type. Alternatively, it can also be utilised as an indicator of a certain numerical value being true, for example, a channel width being greater than 50 cm. The true/false indicator is normally indicated by 1 for true and 0 for false. These values are then used to calculate an experimental semi-variogram which is them modelled and used for a subsequent kriging. The kriging results then provide a probability estimate of the variable under question being true and are within the ranges zero to one. An extension to this method is that of utilising multiple indicators and kriging these thus providing an estimate of the proportional probability of each range.

KRIGING OPTIMISATION

The objective in this section is to do future mining simulations (in the case of large closely spaced follow-up databases where estimation is on an extrapolation basis) or cross-validation (where the primary estimation is of an interpolation nature or data is sparse) to determine:

a. The minimum and maximum number of samples to be used in kriging.
b. The relationship between the kriging regression slope and accuracy of the estimate.
c. The relationship between the KEFF and the accuracy of the estimate or smoothness of the estimate.
d. To what extent one should be using ordinary kriging estimates.

When data is scarce, the process that must be followed must inevitably be one of cross-validation. The cross-validation is done on a point basis with an output file being created of the results. The underlying logic behind the methodology being that if the individual discrete points within the block kriging are of a high quality, the resultant block will also be such. The initial minimum and maximum samples for the cross-validation are set to 2 and an arbitrary high amount such as 60 samples, respectively. This will allow estimates with varying number of samples for subsequent analysis. Both simple and ordinary kriging cross-validation exercises are done with the ordinary kriging results being analysed in the first instance. Here we are attempting to determine the minimum and maximum samples to be allocated for kriging of estimates. The process should follow the following steps which are in a logical order:

a. Run cross-validation exercise as with minimum and maximum samples as indicated previously.
b. Obtain output file and subtract the estimated value from the actual value and divide by the estimated value to obtain the percentage deviation from the actual.
c. Take the absolute value from point (b).
d. Rank the data by number of samples used for estimate.
e. Calculate the average deviation from the estimates per number of samples category.
f. Place data in a graph format and fit a quadratic trend to the data with the number of samples used for the estimate on the x-axis and the average absolute deviation from the actual on the y-axis (Figure 5.23).
g. Differentiate the function to obtain the turning point or simply estimate the lowest point on the graph (see Figure 5.23).
h. Obtain the 10% absolute deviation on either side of the minima. This is your minimum and maximum sample criteria.

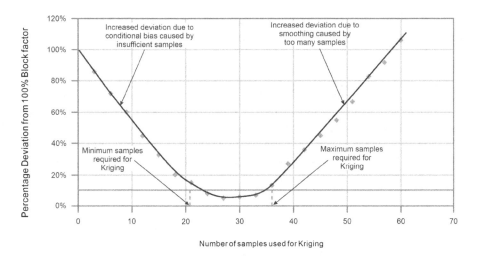

FIGURE 5.23 Example of optimisation of number of samples required for best kriging results.

A slightly different method may be used if sufficient closely spaced follow-up data is available. Here, instead of a cross-validation exercise, one can follow a methodology that is more in line with the method of estimation:

a. Remove a few years' worth of data. This becomes the follow-up data set.
b. Run kriging excluding this data.
c. Bring back the data and superimpose these blocks that have enough samples within the blocks (as per methodology outline elsewhere) over the Kriged model.
d. Obtain the Kriged estimate and follow-up 'actuals'.

From here one follows the same procedure as with the cross-validation from point (b) onwards.

DETERMINING THE KRIGING REGRESSION SLOPE

Using the same estimates versus actual database as in the previous exercise:

a. Using the data obtained from points (a) to (c) in the previous exercise that satisfies the number of samples criteria, now rank the data by kriging regression slope.
b. Calculate the average deviation from the estimates per kriging regression slope categories.
c. Obtain the 10% deviation position as shown in the previous graph. This becomes the ordinary kriging cut-off position.

d. Substitute the simple kriging estimate and simple KEFF for the ordinary kriging estimate for ordinary kriging estimates that fall below the regression slope cut-off.

DETERMINING THE KRIGING EFFICIENCY

This analysis is used in two distinct areas. First, if multiple size grids are being used in the estimation process the kriging efficiency cut-off is used to determine when the grid in question needs to be cut-off to allow the next grid size to take over. Second, in the case where there is only a single block size being estimated, or estimates are extended far beyond data, the analysis serves to inform when the estimates have become smoothed to such an extent that they require some form of post-processing.

> Here the data used is that portion of the data set that satisfies the number of sample criteria as well as having had the substitution of the ordinary kriging by the simple kriging where the kriging regression slope falls below the required level. NB. As the ordinary kriging estimate is being replaced by the simple kriging estimate, due care must be taken to substitute the ordinary kriging efficiency with the kriging efficiency.

As intimated in the previous paragraph, the methodology employed will be determined by numerous factors. These are:

a. Are we trying to determine where we need to make the transition between one block a size and the next?
b. Or are we trying to determine whether post-processing of smoothed estimates is required?

In the case of point (a), it is necessary to decide upon a level of accuracy required from the estimates. This in turn will be used to determine at what stage one needs to cut-off the current estimates and allow the next block size to take over. Under these circumstances, the method is remarkably like that used in the previous two analyses. Here, we plot the absolute deviation of the estimate on the y-axis and the KEFF on the x-axis. The look of the graph will be slightly different under these circumstances with lower kriging efficiencies producing higher deviations, and the percentage deviation will decrease as the KEFF increases. We then simply read off the KEFF equivalent to a 10% absolute deviation or such deviation required, which becomes our estimates' cut-off. That is any estimate that has a KEFF below the figure determined is excluded from the Kriged grid.

In the case of point (b), the analysis described in point (a) is performed but estimates that do not meet the criteria are not removed from the database; instead, these estimates are considered smoothed and thus require post-processing. It is important to note that not all estimates will require post-processing and that only those esti-mates that do not meet the KEFF criteria must receive post-processing.

CASE STUDY: CHANGE OF SUPPORT IN MINERAL RESOURCES ESTIMATION

BACKGROUND

The accuracy and reliability of resource estimates depend on various factors, includ-ing the understanding of the volume–variance relationship. This relationship, which explains the connection between sample size and variability, is key to comprehend-ing the challenges and uncertainties inherent in resource estimation. In this section, we will discuss the concept of change of support and its implications for mineral resources estimation, focusing on the volume–variance relationship as a fundamental principle in this field.

DATA SOURCES AND SCALE CONSIDERATIONS

The primary data used for mineral resource estimation typically come from vertical stratigraphic sections or channels samples, diamond drill core and reverse circulation cuttings. These data represent a relatively small volume compared to the volume rel-evant for mining operations. It is essential to recognise that the observed variability in these small-scale samples will decrease as the volume (support) increases, leading to a more symmetric distribution (Isaaks and Srivastava, 1989). Therefore, a change of support is necessary to bridge the gap between the composited drill hole data and the practical mining scale.

Support refers to the size or volume at which samples are aggregated for estimation purposes and implication for an SMU. An SMU can be defined as the smallest volume that a mining operation can distinguish as either ore or waste (Verly, 2005). The size of the SMU depends on various factors, including the mining method, equipment size, and selectivity characteristics of the deposit. The SMU sizes can range from relatively small dimensions for highly selective operations to larger volumes for bulk-tonnage deposits. As sample size (support) increases, the variability within the grade distribu-tion tends to decrease, and the distribution becomes more symmetric. This is due to the averaging effect of including a mixture of high and low grades within larger sup-port volumes. It is important to note that most grade variables scale linearly, meaning that the mean does not change with support. However, the reduction in variance with increasing support size should be quantified to ensure accurate resource estimation.

The volume–variance effect in the Witwatersrand gold deposits (South Africa) was empirically discovered by Prof. D.G. Krige and it has since become known as the Krige's relationship, or the additivity of variances. The equation for Krige's relation-ship can be expressed as Equation 5.12:

$$\sigma^2\left(\frac{v}{D}\right) = \sigma^2\left(\frac{v}{V}\right) + \sigma^2\left(\frac{V}{D}\right) \tag{5.12}$$

where

- $\sigma^2(v/D)$ represents the variance of the attribute within a small unit (v) or sample relative to the entire deposit or domain (D).
- $\sigma^2(v/V)$ represents the variance of the attribute within the small unit (v) or sample relative to a larger unit (V) or block that contains multiple smaller units or samples.
- $\sigma^2(V/D)$ represents the variance of the attribute within the larger unit (V) or blocks relative to the entire deposit or domain (D).

This equation describes the relationship between the variances at different scales or supports within a mineral deposit. It states that the variance of a small unit (e.g. sample) within the deposit (e.g. domain) is equal to the sum of the variance of that small unit (e.g. sample) within a larger unit (e.g. block) and the variance of these larger units (e.g. blocks) within the entire deposit (e.g. domain). This relationship provides a fundamental understanding of how the variability of mineral attributes changes as the scale or support size varies within a deposit. Using Krige's relationship and examining the volume–variance relationship, mining professionals can gain valuable insights into the distribution and variability of mineral attributes at different scales, facilitating more accurate resource estimation, mine planning, and decision-making processes in the mining industry.

APPLICATIONS AND ANALYSIS OF VOLUME–VARIANCE RELATIONSHIP

To analyse the volume–variance relationship in mineral resource estimation, several practical applications can be explored. These applications provide insights into how the variability of grades changes with different volumes (supports) and offer valuable information for resource estimation and mining operations.

- Sampling design: The volume–variance relationship guides the selection of an appropriate sampling design for grade control purposes. The design should consider the desired support size and the expected reduction in grade variability with increasing support. Through understanding the volume–variance relationship, mining professionals can determine the number and spacing of samples required to accurately estimate the grade within a specific support volume.
- Sampling frequency: In grade control, the frequency of sampling plays a crucial role in capturing the spatial variability of grades. The volume–variance relationship helps in determining the optimal sampling frequency by considering the desired support size and the rate of grade variability reduction. Higher-frequency sampling may be required for smaller support sizes to capture the finer-scale variability, while lower-frequency sampling may be sufficient for larger support sizes.
- Blending piles: Blending piles, also known as homogenisation piles, are used to reduce the variability of material fed into the processing plant. By

blending material from different sources or locations, it is possible to create a more consistent and uniform feed with reduced grade variability. The volume–variance relationship can be employed to assess the effectiveness of blending piles of different sizes and compositions in achieving the desired reduction in variability.

- Selective mining units: SMUs represent the smallest practical volume that can be classified as either ore or waste. Understanding the volume–variance relationship helps in determining the optimal size of SMUs for mining operations. By considering the relationship between support size and grade variability, mining engineers can design selective mining strategies that maximise the extraction of high-grade ore while minimising dilution and waste.

- Grade estimation: The volume–variance relationship influences the grade estimation methods used in grade control. As the support size increases, the grade distribution becomes more symmetric and less variable. Estimation techniques such as inverse distance weighting or ordinary kriging may be suitable for smaller support sizes, while coarser support sizes may require techniques such as inverse kriging or conditional simulation to account for the reduced variability.

- Ore/waste classification: The volume–variance relationship also impacts the classification of material as ore or waste during grade control. Understanding how the grade variability changes with support size helps in defining the cut-off grades or thresholds that distinguish economic material from uneconomic material. The selection of appropriate cut-off grades is crucial for optimising mining operations and minimising the misclassification of material.

Applying these practical applications and analysing the volume–variance relationship, mining professionals can make informed decisions regarding resource estimation, mine planning, and grade control. This understanding helps optimise mining operations, minimise risks associated with grade variability, and maximise the economic value of mineral deposits.

LIMITATIONS AND UNCERTAINTIES

While the volume–variance relationship is a valuable concept in mineral resources estimation and grade control, it is important to acknowledge its limitations and associated uncertainties.

- Scale dependency: The volume–variance relationship is scale-dependent, meaning that the relationship may vary at different scales or supports within a deposit. The observed relationship at one support size may not hold true at another support size. It is crucial to evaluate the volume–variance relationship across multiple scales and understand the scale dependency of grade variability.

- Geological complexity: The volume–variance relationship assumes a certain level of geological homogeneity within the deposit. However, many mineral deposits exhibit geological complexity, including variations in lithology, structure, and mineralisation styles. These complexities can affect the volume–variance relationship and introduce additional uncertainties in resource estimation and grade control.
- Spatial continuity: The volume–variance relationship assumes spatial continuity of grades within the deposit. While geostatistical techniques can help quantify the spatial correlation, there may be areas of the deposit where the continuity breaks down or the grade distribution becomes erratic. In such cases, the volume–variance relationship may not accurately represent the variability of grades.
- Sampling errors: Sampling errors can significantly impact the volume–variance relationship and introduce uncertainties in resource estimation and grade control. The accuracy and reliability of the relationship depend on the representativeness and quality of the collected samples. Sampling errors can arise from various sources, including inadequate sample spacing, sampling bias, improper sample collection techniques, or insufficient sample size. These errors can lead to biased estimates of grade variability and affect the validity of the volume–variance relationship.
- Data quality: The quality of the data used for mineral resource estimation and grade control also influences the volume–variance relationship. Data errors, such as misclassification of samples, incorrect assay results, or incomplete data coverage, can introduce biases and affect the estimation of grade variability. It is essential to implement rigorous data validation and quality control procedures to minimise these errors and ensure the reliability of the volume–variance relationship.
- Uncertainty in estimation methods: The volume–variance relationship is typically incorporated into estimation methods to model grade variability at different support sizes. However, estimation methods themselves introduce uncertainties that can affect the accuracy of the relationship. Estimation parameters, such as the semi-variogram model, interpolation algorithm, or smoothing parameters, can influence the estimated grade variability and, consequently, the volume–variance relationship. It is crucial to assess the uncertainty associated with estimation methods and validate the results against independent data or alternative estimation approaches.
- Temporal variability: The volume–variance relationship is often derived from static data sets representing a specific point in time. However, the grades within a mineral deposit can exhibit temporal variability due to factors such as mining-induced changes, orebody depletion, or natural variations over time. The temporal variability may not be captured adequately by the volume–variance relationship derived from static data. It is essential to consider the temporal aspect and assess the impact of temporal variability on the grade estimation and volume–variance relationship.
- Inherent geological uncertainties: Mineral deposits are inherently complex and subject to geological uncertainties. Variations in mineralisation styles,

geological structures, or alteration patterns can introduce uncertainties in the volume–variance relationship. It is important to incorporate geological understanding, geological modelling, and geospatial analysis techniques to address these uncertainties and refine the volume–variance relationship based on the specific characteristics of the deposit.

Despite these limitations and uncertainties, the volume–variance relationship remains a valuable concept in mineral resources estimation and grade control. It provides a fundamental understanding of how grade variability changes with support size and guides decision-making processes in the mining industry. Mining professionals can implement appropriate mitigation strategies, improve data quality, and refine estimation methods to enhance the accuracy and reliability of resource estimates and grade control practices by acknowledging the limitations and uncertainties associated with the volume–variance relationship.

WORKED EXAMPLE

The case study focuses on the Kloof Reef, which is a conglomerate unit located in the Central Rand Group of South Africa. Specifically, it is situated within the stratigraphy of the 2.90–2.79-billion-year-old rocks. The Kloof Mine, located in the West Rand goldfield, is currently extracting this reef. The Kloof Reef comprises predominantly of interbedded conglomerate and quartzites, and varies in thickness, ranging from 10 cm to over 2 m.

Gold within the Kloof Reef primarily occurs in its native form. The origin of this gold is attributed to fine-grained detrital particles that underwent partial dissolution and subsequent reprecipitation during the infiltration of fluids after the deposition process. This explanation was provided by Frimmel and Nwaila (2020). The dataset for the Kloof Reef consists of 1,464 chip samples obtained from underground stopes and development raise-lines. The dataset includes various fields, as specified in Table 5.2. The spatial distribution of the individual data points can be observed in Figure 5.24. For a comprehensive understanding of the geological context and the gold mineralisation associated with the Witwatersrand Basin, detailed information can be found in the study conducted by Frimmel and Nwaila (2020) (Table 5.4).

After analysing the Kloof Reef database, the next step is to examine the spatial continuity or autocorrelation of the variable of interest, for example, gold grade. One effective tool for this purpose is semi-variogram maps. Semi-variogram maps are graphical representations that display the variation in spatial continuity or autocorrelation of a variable across a geographic area. They are useful tools in geostatistics and spatial analysis for understanding the spatial dependence or structure of a variable. In semi-variogram maps, the semi-variogram values are typically represented using polar coordinates. Each point on the map corresponds to a specific spatial location, and the semi-variogram value at that location indicates the degree of similarity or dissimilarity between pairs of points at different distances and directions. Through examining the semi-variogram values across the map, patterns, and trends in the spatial continuity of the variable can be identified.

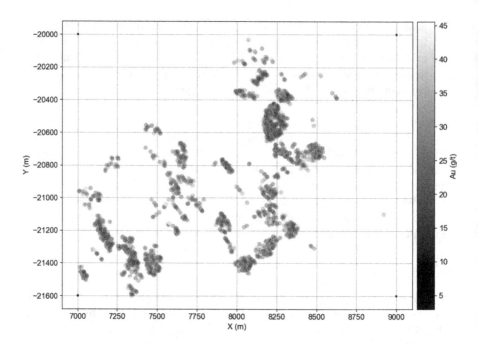

FIGURE 5.24   Plan view showing the distribution of sample localities.

TABLE 5.4

Header structure of the Kloof Reef database

X-Coordinate	Y-Coordinate	Z-Coordinate	Orebody Thickness (cm)	Internal Waste (cm)	% conglomerate	Type of Basal Contact	Au (g/t)	Au (cm.g/t)

Variogram maps can reveal the principal directions or orientations along which the variable exhibits higher spatial continuity or autocorrelation. These principal directions represent the dominant patterns of spatial dependence in the data. Through inspecting semi-variogram maps that include the origin (i.e. semi-variogram values at zero separation distance), one can identify the primary directions along which the variable shows the strongest autocorrelation. The information provided by semi-variogram maps helps in understanding the spatial structure of a variable, identifying trends or anisotropy (directional dependency) in the data, and guiding the selection of appropriate geostatistical models for interpolation or prediction purposes. semi-variogram analysis and semi-variogram maps are widely used in fields such as geology, hydrogeology, geotechnical engineering, environmental sciences, and geostatistics for characterising spatial data and making informed decisions in spatial analysis and modelling (Figure 5.25).

The semi-variogram measures the average dissimilarity between pairs of points as a function of their spatial separation, also known as the lag distance. It quantifies

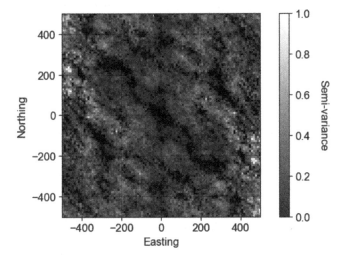

FIGURE 5.25   Semi-variogram map on Cartesian coordinates showing the dominant direction of continuity (continuous darker colours at the centre of the plot).

the spatial variability and provides insights into the correlation structure of the variable being analysed. Semi-variograms are useful tools in spatial analysis for characterising the spatial dependence or variability of a variable across a study area. They provide insights into the patterns and structures of spatial correlation, which can be valuable for various applications such as spatial interpolation, geostatistical modelling, and resource estimation. The main utility of experimental semi-variograms lies in their ability to quantify the spatial autocorrelation of a variable at different distances and directions. Through the calculation of the semi-variogram, which measures dissimilarity or variability between pairs of sample points across different lags or separation distances, a semi-variogram is formed. It is important to differentiate different structural behaviours of variograms. Differentiating between isotropic and anisotropic semi-variograms is done by observing the spatial variability in the graphical representation of the experimental semi-variogram. Here is how it can be done:

a. Isotropic Semi-variogram: In an isotropic semi-variogram, the spatial variability is similar in all directions. This is typically indicated by a smooth and symmetric semi-variogram curve. The values of semi-variogram increase with increasing distance up to a certain point, known as the range, beyond which the semi-variogram remains relatively constant. The range represents the distance beyond which spatial dependence is negligible or insignificant.

b. Anisotropic Semi-variogram: In an anisotropic semi-variogram, the spatial variability varies across different directions. This is reflected in the shape of the semi-variogram curve, which may be elongated or asymmetrical. The semi-variogram values may vary depending on the direction or orientation of the lag vector. Anisotropic semi-variograms indicate that there is a preferential direction or spatial anisotropy in the data, suggesting that the spatial dependence is stronger or weaker in specific directions (Olea, 2012).

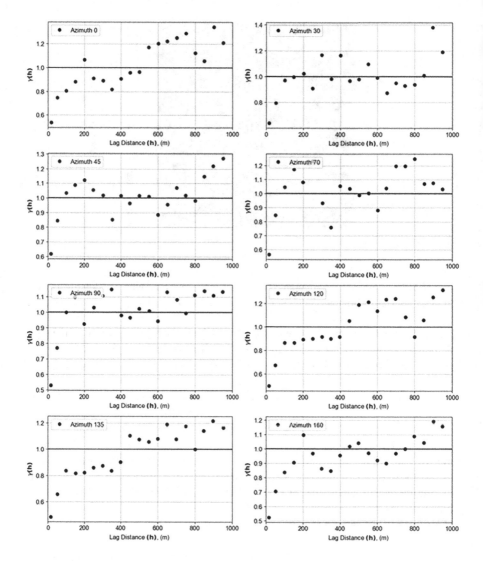

FIGURE 5.26 Experimental semi-variograms showing variation in spatial continuity.

Once the semi-variogram has been calculated, positive-definite model is fitted (Figure 5.26). In this case study example, the spherical model which is a commonly used mathematical model to fit the experimental semi-variogram in most metal data was applied. To fit a spherical model to a semi-variogram, several parameters need to be estimated: the nugget effect, the sill, and the range.

- Nugget Effect: The nugget effect (C_0) represents the discontinuity or variability at very short distances (lag = 0). It accounts for measurement errors, sampling errors, or small-scale variability that cannot be captured by the model. It is the intercept of the semi-variogram at the origin.

- Sill: The sill (C) represents the maximum variability or range of spatial correlation in the data. It is the semi-variogram value when the lag distance tends to infinity. It indicates the overall level of spatial dependence or autocorrelation.
- Range: The range (a) is the distance at which the spatial correlation or dependence becomes negligible. Beyond this distance, the semi-variogram reaches a plateau or levels off, indicating that the points are no longer correlated.

The parameters of the spherical model (C_o, C, a) are estimated by fitting the model to the empirical semi-variogram derived from the data (Olea, 1991). This involves minimising the difference between the model's values and the observed semi-variogram values at different lags. The fitting can be done using various optimisation techniques such as least squares or maximum likelihood estimation.

Modelling anisotropy

Anisotropy refers to the directional dependence of spatial variability. In many geostatistical applications, spatial processes exhibit different levels of correlation or variability in different directions. To account for anisotropy, an anisotropic semi-variogram model is used instead of the isotropic spherical model. Anisotropic models introduce additional parameters that describe the orientation and anisotropy ratio (or anisotropy factor) of the spatial dependence. The anisotropy ratio indicates how the spatial correlation varies between different directions. The range parameter is modified to represent the correlation range in each direction separately. To model anisotropy, the orientation and anisotropy ratio need to be estimated along with the nugget effect, sill, and range parameters. This can be done by rotating the coordinate system and fitting an appropriate anisotropic model (e.g. anisotropic spherical model) to the semi-variogram in the rotated directions. The rotation and estimation process can be performed iteratively until an optimal fit is achieved. The generated scatter points and the fitted spherical semi-variogram models are visualised using a plot. The x-axis represents the lag distance in metres, and the y-axis represents the semi-variance. The major direction scatter points are represented by blue dots, while the minor direction scatter points are represented by red dots. The major and minor direction models are plotted as continuous lines. The green dotted line indicates the sill, which represents the maximum semi-variance value.

Importance in kriging

The fitted anisotropic spherical semi-variogram is used in the kriging process in the following ways:

- Spatial Interpolation: Kriging uses the spatial correlation structure revealed by the semi-variogram to interpolate values at unobserved locations (Isaaks and Srivastava, 1988). By fitting the semi-variogram with the spherical model, we can estimate the spatial dependence and variability of the variable of interest. This information is then used to predict values at unsampled locations more accurately.

- Optimal weighting: Kriging assigns weights to the observed data points based on their spatial proximity and similarity to the unsampled location being estimated. The weights are determined using the semi-variogram, where points that are spatially closer and exhibit higher correlation have higher weights. The anisotropic nature of the spherical semi-variogram accounts for directional variations in the spatial dependence, allowing for more precise and accurate predictions.

- Error Estimation: Kriging provides not only point estimates but also estimates of the prediction uncertainty. The semi-variogram is instrumental in quantifying the prediction error by providing an indication of the spatial correlation between data points. The fitted spherical model aids in characterising the spatial variability, allowing for better estimation of the prediction uncertainty (Isaaks, 2005).

- Optimal sampling design: The semi-variogram can also guide the design of optimal sampling schemes. Examining the spatial dependence structure as indicated by the semi-variogram allows for the identification of regions

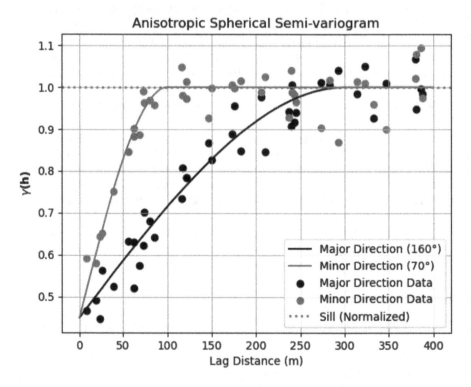

FIGURE 5.27 The horizontal major (045°) and minor (125°) directions of continuity on the X–Y plane.

exhibiting either elevated uncertainty or diminished spatial correlation. This information can inform the selection of additional sampling locations to improve the accuracy and efficiency of the kriging estimates (Figure 5.27).

Micro kriging and macro kriging are two important techniques in geostatistics that are important in grade control and mine planning processes (Figure 5.28). They are used to model and estimate the distribution of valuable minerals or metals within a mining deposit. Micro kriging focuses on smaller-scale variability and is primarily used for grade control purposes. It involves estimating the grades of individual smaller blocks or samples within the deposit. This level of estimation provides detailed information about local variations in mineral grades, which is important for short-term mine operations and immediate production decisions.

On the other hand, macro kriging is applied to larger blocks and considers broader-scale variability in the deposit (Figures 5.29 and 5.30). It is used for mine planning at different time scales. The estimates generated through macro kriging are used for short-term mine planning, medium-term business planning (Figure 5.30), and long-term life-of-mine planning (Figure 5.30). Using macro kriging, mining companies can assess the overall mineral distribution, identify major trends and

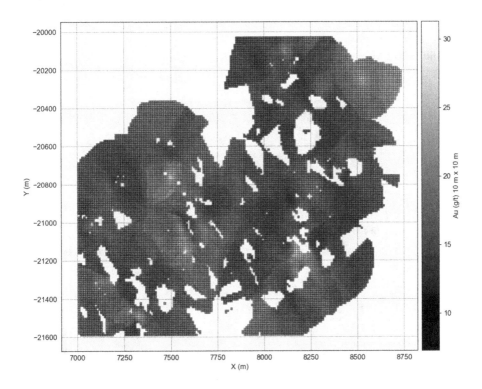

FIGURE 5.28 Micro kriging grade control block model.

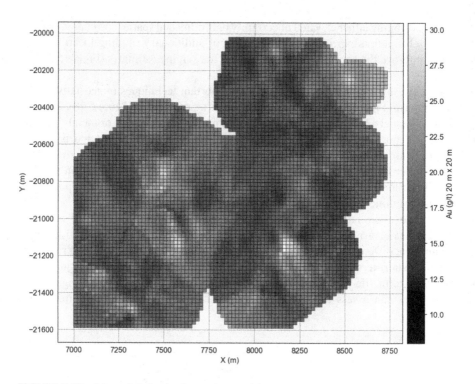

FIGURE 5.29   Macro kriging business plan model.

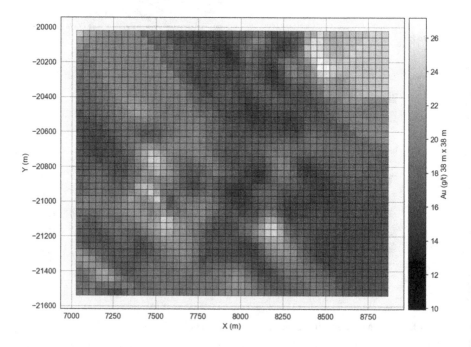

FIGURE 5.30   Macro kriging life-of-mine block model.

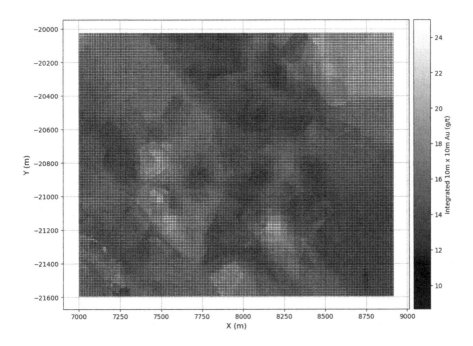

FIGURE 5.31 Integrated mineral resources model combining short-, medium-, and long-term estimates.

structures, and make informed decisions about resource allocation and mine development strategies.

The integration of micro and macro blocks into a single model allows for a more realistic representation of the mineral distribution within the SMU (Figure 5.31). In this context, it is important to ensure that the micro blocks override the macro blocks rather than the other way around. This means that the estimation results from micro kriging should take precedence over macro kriging when considering local-scale variations. This is because grade control decisions and short-term mine operations require accurate information about the immediate area being mined, and micro kriging provides the necessary level of detail for these purposes.

CONCLUSION

In conclusion, geostatistics grapples with a significant challenge of assimilating a wide array of data sources into a coherent numerical geological model. These sources encompass various measurements, both direct and indirect, along with analogue information, necessitating a rigorous approach to manage the substantial volumes of data that are intrinsic to geostatistical investigations. Embracing the data event methodology provides an opportunity to enhance data management, but exercising utmost care during data preparation and cleansing is crucial to prevent any adverse impacts on model quality and consequential decisions. The process of geostatistical resource modelling follows a logical methodological sequence of data collection, analysis, and inference. The initial steps in mineral resource estimation involve inputting samples and measurements, processing the data through compositing and applying high-grade cut-offs. This leads to the construction of the geological model by defining geological zones and estimation domains and wireframing mineralisation and related features, as well as characterising contacts and spatial distribution. Additionally, structural analysis is carried out by coordinating transformations, while rock density measurements and spatial distribution models of rock densities are established.

To gain a deeper understanding, EDA aids in exploring descriptive statistics within domains, assessing population homogeneity, identifying outliers, examining attribute correlations, declustering data, and detecting spatial trends and discontinuities. Geostatistical analysis, including variography, is then employed to investigate spatial data autocorrelations, apply transformations to enhance structures, and test statistical hypotheses and geostatistical assumptions. The critical next step lies in selecting appropriate estimation methods and defining modelling parameters, such as search neighbourhoods and block sizes. This phase involves implementing the chosen modelling algorithm through computer scripts and macro codes, accompanied by thorough validation and sensitivity testing of the estimate to ensure its reliability. Subsequently, the classification of mineral resource categories is undertaken, followed by comprehensive documentation, including tabulations of resource grade and tonnage, along with a clear summary of the employed data and methodologies. Executing these preliminary steps effectively guarantees a robust and accurate foundation for subsequent resource management decisions. By amalgamating a solid geological understanding, rigorous data handling, and advanced geostatistical techniques, this process yields valuable insights into the distribution and quality of mineralisation, enabling informed decision-making and optimisation of mining operations. The indispensable collaboration between geologists and statisticians remains at the core of this process, culminating in reliable resource estimates that enhance the productivity and sustainability of the mining industry.

NOTES

1　In stratigraphic formations, multiple layers of sedimentary rock, soil, or igneous rocks are laid in parallel by natural processes. These layers possess horizontal extent but exhibit a smaller vertical dimension upon deposition. However, folding and faulting can alter the original continuity, leading to the distorted layers observed today. To effectively characterize such formations for boundary and subsurface property modelling (encompassing petrophysical properties, facies, rock types, and mineral grades), it is crucial to model the non-linear spatial continuity along these layers.

Fortunately, the distortion of the original strata can be reversed through the utilisation of modelling techniques that consider bounding surfaces and the correction of sample and grid coordinates within the strata. This corrective process creates a suitable space for applying linear geostatistical techniques such as kriging for accurate modelling. Such a coordinate transformation becomes necessary and applicable for all distorted yet essentially planar formations, including extensive veins occurring within a single plane. By employing these methodologies, we can gain a comprehensive understanding of stratigraphic deposits and their intricate characteristics in subsurface environments.

2 A **t-test** is a type of inferential statistic used to determine if there is a significant difference between the means of two groups, which may be related in certain features. It's particularly useful in hypothesis testing to help interpret whether differences in data samples could have arisen by chance, or whether they reflect true differences in the underlying populationsReferences

REFERENCES

Abzalov, M. Z. (2006). Localised uniform conditioning (LUC): A new approach for direct modelling of small blocks. *Mathematical Geology, 38,* 393–411.

Abzalov, M. (2008). Quality control of assay data: A review of procedures for measuring and monitoring precision and accuracy. *Exploration and Mining Geology, 17*(3–4), 131–144.

Abzalov, M. Z. (2009). Use of twinned drill – holes in mineral resource estimation. *Exploration & Mining Geology, 18*(1–4), 13–23.

Abzalov, M. (2011). *Sampling errors and control of assay data quality in exploration and mining geology* (pp. 611–644). IntechOpen.

Abzalov, M. (2016). Methodology of the mineral resource classification. In: *Applied Mining Geology. Modern Approaches in Solid Earth Sciences,* Vol. 12, 355–363. Springer, Cham. https://doi.org/10.1007/978-3-319-39264-6_28.

Deutsch, J. L. & Deutsch, C. V. (2012). A new version of kt3d with test cases. *Centre for Computational Geostatistics Annual Report, 14,* 403–410.

Deutsch, J. L., Szymanski, J., & Deutsch, C. V. (2014). Checks and measures of performance for kriging estimates. *The Journal of the Southern African Institute of Mining and Metallurgy, 114,* 223–230.

Dias, P. M., & Deutsch, C. V. (2022). The decision of stationarity. In J.L. Deutsch (Ed.), *Geostatistics lessons.* Retrieved from http://www.geostatisticslessons.com/lessons/stationarity

Frimmel, H. E. & Nwaila, G. T. (2020). Geologic evidence of syngenetic gold in the Witwatersrand Goldfields, South Africa. In Sillitoe, R. H., Goldfarb, R. J., Robert, F., & Simmons, S. F. (Eds.), *Geology of the world's major gold deposits and provinces.* Society of Economic Geologists.

Isaaks, E. (2005). The Kriging Oxymoron: A conditionally unbiased and accurate predictor (2nd Edition), Quantitative Geology and Geostatistics, Geostatistics 2004, Banff, Volume 14 (pp. 363–374).

Isaaks, E. H. & Srivastava, R. M. (1988). Spatial continuity measures for probabilistic and deterministic geostatistics. *Mathematical Geology, 20,* 313–341.

Isaaks E. H. & Srivastava R. M. (1989). *An introduction to applied geostatistics.* New York: Oxford University Press.

Krige, D. G. (1951). A statistical approach to some basic mine valuation problems on the Witwatersrand. *Journal of the Chemical, Metallurgical and Mining Society of South Africa, 52,* 119–139.

Krige, D. G. (1997). A practical analysis of the effects of spatial structure and of data available and accessed, on conditional biases in ordinary kriging. In Baafi, E. Y. & Schofield, N. A. (Eds). *Geostatistics. Wollongong '96, Fifth International Geostatistics Congress, Wollongong, Australia,* September 1996 (Vol. 1, pp. 799–810). Dordrecht: Kluwer.

Larrondo, P. & Deutsch, C. V. (2005). Accounting for geological boundaries in geostatical modeling of multiple rock types. In *Geostatistics, Banff 2004* (pp. 3–12). Dordrecht: Springer Netherlands.

Manchuk, J. (2010). *Geostatistical modeling of unstructured grids for flow simulation (PhD thesis).* University of Alberta, Edmonton, Canada.

Olea, R. A. (1991). *Geostatistical glossary and multilingual dictionary*. New York: Oxford University Press.

Olea, R.A. (1999). Ordinary Kriging. In *Geostatistics for engineers and earth scientists* (pp. 39–65). Boston, MA: Springer. https://doi.org/10.1007/978-1-4615-5001-3_4. https://link.springer.com/book/10.1007/978-1-4615-5001-3.

Olea, R. A. (2012). *Geostatistics for engineers and earth scientists*. Springer Science & Business Media.

Shaw, W. J. (1997). *Validation of sampling and assaying quality for bankable feasibility studies*. The Resource Database Towards 2000, Melbourne: AusIMM, pp. 69–79.

Sinclair, A. J. & Bentzen, A. (1998). Evaluation of errors in paired analytical data by a linear model. In Vallée, M., and A. J. Sinclair (Eds.), *Quality assurance, continuous quality improvement and standards in mineral resource estimation* (Vol. 7, nos. 1 and 2, pp. 167–174), *Expl. and Min. Geol.* Cambridge University Press.

Sinclair, A. J. & Blackwell, G. H. (2006). *Applied mineral inventory estimation*. Cambridge University Press.

Snowden, D. V. (2001). Practical interpretation of mineral resource and ore reserve classification guidelines. In Edwards, A. C. (Ed.), *Mineral resource and ore reserve estimation – The Australasian Institute of Mining and Metallurgy (AusIMM)* (pp. 643–652). Guide to Good Practice.

Stanley, C. R. & Lawie, D. (2007). Average relative error in geochemical determinations: Clarification, calculation, and a plea for consistency. *Exploration and Mining Geology, 16*(3–4), 267–275.

Verly, G. (2005). Grade control classification of ore and waste: A critical review of estimation and simulation based procedures. *Math Geol, 37*, 451–475. https://doi.org/10.1007/s11004-005-6660-9

Yunsel, T. Y. & Ersoy, A. (2011). Geological modeling of gold deposit based on grade domaining using plurigaussian simulation technique. *Natural Resources Research, 20*, 231–249.

6 Mineral resources and mineral reserves estimation process

INTRODUCTION

The process of estimating mineral resources and mineral reserves is a critical task of the mining industry. This process involves assessing and modelling the grade, tonnage, and location of subsurface mineral deposits. Due to the high costs associated with sampling, especially in underground environments and in the solid rock context, the sampling budget typically limits the amount of data available for mineral resource estimation. As a result, extracting as much information as possible using replicable and reliable established methods is important to ensure that the resulting estimates are as accurate and precise as possible (Abzalov, 2013). Estimating mineral resources using geostatistics involves assigning weights to samples based on their informational value. This is typically done by identifying the grade continuity of the deposit, which refers to how grades change across space. There are different approaches to estimation, depending on the specific objectives and constraints of the mining operation.

The scale of the estimation varies depending on the specific task being accomplished. There are two main scales of estimation based on typical mine operations which are deposit and local scales (Annels, 1997). For long-term mine planning or life-of-mine estimation, the deposit scale approach is used, because the coverage must encompass the relevant portions of the deposit for strategic planning. For this purpose, widely spaced samples from drilling campaigns are used to accurately define the global grade-tonnage curve (Table 6.1). This helps to ensure that the overall resource estimate is unbiased and reflects the full extent of the deposit. However, because the samples are widely spaced, the resulting estimate may be biased locally and may not accurately reflect the variability of grades within the deposit (Abzalov, 2013; Annels and Dominy, 2002). In contrast, local scale estimation is required for tactical and operational purposes, typically grade control. This process results in a short-term estimation curve (Table 6.1) for the purpose of ore grade-control and requires higher resolution drilling and sampling to accurately define grades at a local scale, typically at the scale of a selective mining unit (SMU). The local scale is used to define the ore-waste boundary locally and ensures that only ore is extracted from the deposit. The resulting estimate is highly localised and captures, as much as possible, the variability of grades within the deposit at the scale of extraction.

Regardless of the approach used, mineral resource estimation is inherently uncertain due to the limited amount of data available. To account for this uncertainty, it

DOI: 10.1201/9781032650388-6

TABLE 6.1

Comparison of long- and short-term resource estimation and associated criterion

Criterium	Long-Term	Short-Term
Block size	Variable; depends on many factors	SMU size
Minimum and maximum number of samples	Often restricted to preserve variability	A large number of samples
Number of samples from an individual drill hole	Enforce >1 drill hole for estimation	Usually, enough data to not be an issue
Search distances and orientations	Geology and grade continuity controlled	Geology and grade continuity controlled
Estimation technique	Inverse Distance, Kriging, Change-of-Support techniques (i.e., Block kriging, Multiple Indicator Kriging, Uniform Conditioning, Conditional Simulation)	Kriging, Conditional Simulation

is important to use statistical methods and assess the sensitivity of the estimates to different assumptions and inputs. Overall, the process of mineral resource estimation is complex and challenging, but it is critical for making informed decisions about the development and operation of a mine. By using robust estimation methods and accounting for uncertainty, mining companies can ensure that they are making the best use of their resources while minimising the risk of over- or underestimating the value of the deposit.

Various methods are available for estimating the grade and tonnage of mineral deposits, each with its own set of advantages and disadvantages. Regardless of the approach, a reliable estimation requires a robust and high-quality database of samples, a sound geological model, and meticulous validation.

LINEAR INTERPOLATION TECHNIQUES

a. Radial Basis Functions

Radial Basis Function (RBF) interpolation is a mathematical technique used for approximating or interpolating data points in multidimensional space. It is commonly employed in various fields, including geology, computer graphics, machine learning, and engineering. RBF interpolation works by using a set of basis functions to interpolate data points based on their distances from specific centres. Here is a general overview of how RBF interpolation methods work:

i. RBF: A RBF is a mathematical function that depends only on the distance from a centre or reference point. The most commonly used RBF is the Gaussian function:

$$\phi(r) = \exp\left(-\varepsilon \times r^2\right)$$

where r is the distance between the data point and the centre, and ε is a parameter that controls the shape and smoothness of the interpolation.

Other RBF functions include:

- Linear = $-r$
- Thin plate spline = $r^2 \times \log(r)$
- Cubic = r^3
- Quintic = $-r^5$
- Multiquadric = $-\text{sqrt}(1 + r^2)$
- Inverse multiquadric = $1/\text{sqrt}(1 + r^2)$
- Inverse quadratic = $1/(1 + r^2)$

ii. Data points and centres: In RBF interpolation, you have a set of data points with known values and their respective coordinates in multidimensional space. The centres are also points in the same multidimensional space and act as reference points for interpolation.

iii. Weighted sum: To interpolate a new data point, the RBF method computes the weighted sum of the values of the known data points, where the distances between the new point and the centres determine the weights. Closer data points have more influence on the interpolation than distant ones.

iv. Solving the system: The process involves solving a system of linear equations to determine the appropriate weights for each data point in the interpolation process. This system is constructed based on the distances between the data points and the centres and the target function values.

v. Choosing centres: Selecting the right centres is crucial for an accurate interpolation. They can be chosen as the same points as the data points, or they can be distributed differently (e.g., using k-means clustering).

vi. Parameter selection: The parameter ε in the Gaussian function controls the smoothness of the interpolation. It needs to be chosen appropriately, as a high ε value may lead to an over-smoothed surface, while a low ε value may cause overfitting.

RBF interpolation methods offer advantages such as flexibility, ease of implementation, and the ability to handle data scattered irregularly in space. However, their performance can be sensitive to the proper selection of centres and parameters. It is important to note that there are other types of RBFs and variations of the method used in different applications, but the fundamental principle remains the same: approximating data points based on distances from centres using radial basis functions.

b. Triangulation

Triangulation is another interpolation method commonly used in various fields, including geology, computer graphics, and geographic information systems (GIS). Unlike RBF interpolation, triangulation uses a different approach to approximate or interpolate data points in 2D or 3D space. Triangulation is particularly useful when working with scattered or unstructured data points. Here is a general overview of how triangulation interpolation works:

i. Triangulation: In triangulation, the data points are connected to form a network of triangles. Three non-collinear data points define each triangle, and no data points lie inside the triangles. This network of triangles is often referred to as a 'Delaunay triangulation', which ensures that no data points are inside the circumcircles of the triangles.

ii. Interpolation within Triangles: Once the triangulation is established, interpolating a new data point involves finding the triangle containing the new point. The value of the new point is then interpolated based on the values of the three vertices of the containing triangle.

iii. Barycentric coordinates: Barycentric coordinates are used to perform the interpolation within a triangle. Barycentric coordinates express the position of a point within a triangle as a weighted sum of the triangle's vertices. The weights are determined by the relative areas of sub-triangles formed by the new point and the three vertices of the containing triangle.

iv. Linear Interpolation: Once the barycentric coordinates are determined, linear interpolation is used to compute the value of the new point based on the values of the three vertices. The interpolated value is a weighted sum of the vertex values, with weights given by the barycentric coordinates.

v. Triangulation Algorithms: Various algorithms exist to perform triangulation efficiently. One of the most common algorithms is the 'Bowyer–Watson algorithm', which iteratively inserts the data points one by one to construct the Delaunay triangulation.

Triangulation has several advantages, such as simplicity, efficiency, and the ability to handle irregularly spaced data points. However, it may have limitations when dealing with data points close to the boundary of the data set, as there might not be enough neighbouring points to form valid triangles. In such cases, additional techniques such as extrapolation or using a boundary condition may be necessary.

c. Inverse Distance Weighting

Inverse Distance Weighting (IDW) is another popular interpolation method used in geology, environmental science, GIS, and various other fields. Like RBF and triangulation, IDW is used to estimate values at unknown locations based on known values at nearby locations. IDW is a deterministic interpolation method that assigns weights to nearby data points inversely proportional to their distances from the target location. Here is a general overview of how IDW works:

i. Distance calculation: For a given target location where we want to interpolate a value, IDW calculates the distances between the target location and all the known data points (sample points). The distance can be

measured using Euclidean distance or other distance metrics, depending on the application.

ii. Weight calculation: After calculating the distances, IDW assigns
weights to each data point based on their inverse distances from the
target location. The closer data points are given higher weights, while
farther data points receive lower weights. The formula for calculating
the weight (W) for each data point (i) is typically given by:

$$W_i = \frac{1}{(\text{distance}_i)^p}$$

where 'distance$_i$' is the distance between the target location and the ith
data point, and p is a positive power parameter. The choice of the power
parameter affects the influence of distant points on the interpolation.
A high p value gives more weight to closer points, while a low p value
spreads the influence of points more evenly.

iii. Interpolation: Once the weights are calculated, IDW performs a
weighted average of the known data points' values, using the weights as
the weighting factors:

$$\text{Interpolated value at the target location} = \frac{\Sigma(W_i \times \text{value}_i)}{\Sigma W_i}$$

where the summation is performed over all data points, and 'value$_i$' is
the value of the ith data point.

iv. Cell size and search radius: In practice, a cell size or search radius may
be defined to control the spatial resolution of the interpolation and limit
the number of data points considered for interpolation. The interpolation will only consider the data points within this radius.

IDW is relatively straightforward to implement and interpret. However,
its performance can be sensitive to the power parameter (p) and the search
radius or cell size choice. As with any interpolation method, it is crucial to
consider the characteristics of the data, the spatial distribution of sample
points, and the application's specific requirements when using IDW. IDW
should be avoided whenever possible, this is because, very rarely does the
actual spatial relationship correspond to an inverse distance model.

d. Nearest Neighbour

Nearest Neighbour interpolation is one of the simplest and most straightforward interpolation methods used in various fields, including image processing, computer graphics, and GIS. It is a non-parametric method that
estimates the value of a target location based on the value of the nearest
known data point (sample point). Here is a general overview of how Nearest
Neighbour interpolation works:

i. Data points:

In Nearest Neighbour interpolation, you have a set of known data
points with their respective values and coordinates in a 2D or 3D space.

ii. Interpolation:

Nearest Neighbour interpolation identifies the closest known data point to the target location based on Euclidean distance or other distance metrics when you want to interpolate a value at a target location.

iii. Value assignment:

The value of the target location is then assigned the same value as the nearest data point. In other words, the interpolated value is simply the value of the nearest data point.

iv. No weighting:

Unlike other interpolation methods such as IDW or RBF, Nearest Neighbour interpolation does not involve weighting or averaging neighbouring data points. The value at the target location is directly copied from the closest data point.

v. Discrete grids:

Nearest Neighbour interpolation is particularly useful when dealing with discrete data grids, such as pixel-based images or raster datasets. It is commonly used for tasks such as image resizing or resampling, where new pixel values need to be determined based on the original pixel values.

vi. Advantages and limitations:

The main advantage of Nearest Neighbour interpolation is its simplicity and efficiency. It is computationally straightforward, making it quick and easy to implement. Additionally, it preserves the exact values of the known data points, making it useful when you need to preserve sharp boundaries or discrete data. However, Nearest Neighbour interpolation has some limitations. One major limitation is its tendency to produce a 'blocky' or 'stair-step' effect, primarily when used for image resizing, as it does not consider the values of neighbouring points. This can lead to visible artefacts in the interpolated result in areas with varying data. Due to its simplicity and limitations, Nearest Neighbour interpolation is often used in cases where speed and simplicity are more critical than achieving smooth and accurate interpolation results. In applications with higher accuracy and smoother results, other interpolation methods such as bilinear or bicubic interpolation are often preferred.

e. Simple Kriging

Simple Kriging is a geostatistical interpolation method used in geology, environmental science, and other fields to estimate values at unsampled locations based on a spatially correlated dataset. It is one of the variants of Kriging, a widely used set of geostatistical techniques for spatial interpolation. Kriging methods consider both the spatial structure of the data and the uncertainty associated with the estimation. Here is a general overview of how Simple Kriging works:

i. Data points and spatial variability: In Simple Kriging, you have a set of known data points with their respective values and spatial coordinates. The underlying assumption is that the variable being interpolated (e.g., soil contamination level or mineral concentration) exhibits spatial variability, meaning that its values are correlated with the spatial distance between data points.

ii. Variogram calculation: The first step in Simple Kriging is to calculate the variogram or semi-variogram of the dataset. The variogram measures the spatial correlation between data points at different distances. It characterises how the variability of the variable of interest changes with distance. The variogram is used to model the data's spatial autocorrelation and anisotropy (directional dependence).

iii. Variogram modelling: The next step is to fit a mathematical model to the experimental variogram. The model parameters define the variogram's range, sill, and nugget. The range represents the distance beyond which data points are no longer spatially correlated. The sill represents the total variance of the data, and the nugget represents the nugget effect, which accounts for spatial variability at very short distances (measurement errors and microscale variation).

iv. Kriging weight calculation: Once the variogram model is fitted, Simple Kriging calculates the spatial weights (e.g. Kriging weights) for the surrounding data points based on their distances and the variogram model. The weights are determined in such a way that they minimise the estimation error and account for the spatial correlation of the variable.

v. Interpolation: The interpolated value at the unsampled location is then obtained as a weighted sum of the values of the surrounding data points using the Kriging weights. The interpolation takes into account the spatial autocorrelation and the uncertainty associated with the estimation.

vi. Kriging variance: In addition to the interpolated value, Simple Kriging provides an estimate of the estimation variance (Kriging variance). This variance quantifies the uncertainty associated with the interpolation at the unsampled location. It is useful for assessing the reliability of the interpolated values.

Simple Kriging is well-suited for cases where the underlying spatial structure of the data is known to be stationary (homogeneous) and where the variogram exhibits a well-defined pattern. It provides accurate interpolations and produces optimal linear unbiased estimates (BLUE) under certain assumptions. However, it requires the determination of the variogram model, which can be challenging in practice, especially for datasets with complex spatial patterns or anisotropy.

f. Ordinary Kriging

Ordinary Kriging is another geostatistical interpolation method, such as Simple Kriging, used to estimate values at unsampled locations in geology, environmental science, and various spatial analysis applications (Krige, 1996). Like other Kriging methods, Ordinary Kriging considers the spatial structure and variability of the data to provide optimal estimates and quantify uncertainty (Deutsch and Journel, 1998). Here is a general overview of how Ordinary Kriging works:

i. Data Points and Spatial Variability: Similar to Simple Kriging, in Ordinary Kriging, you have a set of known data points with their respective values and spatial coordinates. The interpolated variable is assumed to have spatial variability, and its values correlate with the spatial distance between data points.

ii. Variogram calculation and modelling: The first step in Ordinary Kriging is to calculate the variogram or semi-variogram of the dataset, just like in Simple Kriging. The variogram measures the spatial correlation between data points at different distances. A variogram model is then fitted to the experimental variogram to characterise the data's spatial autocorrelation and anisotropy (directional dependence).

iii. Kriging Weight Calculation: In Ordinary Kriging, the Kriging weights are calculated for the surrounding data points based on their distances and the variogram model. The Kriging weights are determined in such a way that they provide the BLUE of the variable at the unsampled location. Unlike Simple Kriging, Ordinary Kriging assumes that the mean of the variable is unknown and not fixed, which means it allows for trend estimation. To calculate the Kriging weights (λ_i) that minimise the variance of the estimation error while incorporating the trend analysis, we introduce a constraint. The constraint is that the sum of the Kriging weights should be equal to one ($\Sigma \lambda_i = 1$). This ensures that the interpolation is unbiased, meaning that the Kriging estimate is not biased towards any data point. Mathematically, we want to minimise the following objective function: $F(\lambda_1, \lambda_2, ..., \lambda_n) = $ Kriging variance + $\lambda \times (\Sigma \lambda_i - 1)$

where 'Kriging variance' represents the estimation error variance, and λ is the Lagrange multiplier associated with the constraint. By taking the partial derivatives of F with respect to each λ_i and λ, and setting them to zero, we can find the values of λ_i and λ that satisfy the constraint and minimise the Kriging variance. Once the Lagrange multipliers are determined, the Kriging weights can be calculated, and the final interpolated value at the unsampled location is obtained as the sum of the local mean (trend) and the weighted sum of the values of the surrounding data points.

iv. Trend Analysis: Ordinary Kriging incorporates trend analysis to estimate the local mean of the variable. The trend represents any systematic spatial variation in the variable that cannot be explained by the spatial autocorrelation modelled by the variogram. The trend is typically modelled using a mathematical function (e.g. linear and polynomial) and estimated as part of the interpolation process.

v. Interpolation: The interpolated value at the unsampled location is obtained as the sum of the local mean (trend) and the weighted sum of the values of the surrounding data points using the Kriging weights. The interpolation accounts for both the spatial autocorrelation captured by the variogram and the local trend.

vi. Kriging Variance: Similar to Simple Kriging, Ordinary Kriging provides an estimate of the estimation variance (Kriging variance). This variance quantifies the uncertainty associated with the interpolation at the unsampled location. It considers both the spatial variability captured by the variogram and the uncertainty introduced by the local trend estimation.

NONLINEAR TECHNIQUES FOR GRADE AND TONNAGE ESTIMATION

- Uniform Conditioning: This method involves conditioning the estimation on a set of auxiliary variables, such as lithology or structural data, to improve the accuracy of the estimate. This approach can be highly effective when there is a strong relationship between the auxiliary variables and the grade and tonnage distribution.
- Multiple Indicator Kriging: This technique is similar to Uniform Conditioning but involves the use of multiple indicator variables to better capture the geological complexity of the deposit. This approach is highly effective when there are multiple geological domains within the deposit.
- Indicator Kriging: This method is similar to Ordinary Kriging but takes into account the binary nature of the deposit, such as whether a sample point is above or below a particular cut-off grade. This approach can be useful when the deposit has a discrete nature, and the estimation must account for the presence or absence of the mineralisation.
- Sequential Gaussian Simulation: This technique involves simulating multiple possible realisations of the grade and tonnage distribution while honouring the available data and spatial correlation. The simulations are generated sequentially, with each new realisation conditioned on the previous one. This approach can capture the uncertainty associated with the estimation process and provide a range of possible values.
- Multiple-Point Geostatistics: This method is similar to conditional simulation but uses a more complex approach to simulate the grade and tonnage distribution. Multiple-Point Geostatistics uses a training image to guide the simulation and can capture complex geological features such as faults and channels.
- Support Vector Machines: This approach involves using machine learning algorithms to estimate the grade and tonnage distribution based on the available data. Support Vector Machines can handle high-dimensional data and can capture non-linear relationships between the data and the target variable.
- Artificial Neural Networks: This technique also uses machine learning algorithms to estimate the grade and tonnage distribution based on the available data. Artificial Neural Networks can capture complex relationships between the data and the target variable and can handle noisy data.

The selection of the appropriate method for grade and tonnage estimation depends on a range of factors, including the quality and quantity of available data, the geological complexity of the deposit, and the desired level of accuracy and uncertainty (Annels and Dominy, 2002). Therefore, it is essential to carefully evaluate the available methods and select the one that is best suited for a given situation.

PURPOSE AND APPROACH

As the reader may have already become aware of estimation utilising Kriging is not merely a matter of calculating an experimental semi-variogram, modelling it and then doing a Kriging run. Firstly, one needs to determine some basics, which in turn will result in different methods being employed to obtain the prerequisite result (Krige, 1996). These are:

a. What are we going to be using the Kriging for?
   i.   Monthly planning.
   ii. Monthly estimation of results.
   iii. Medium-term planning, such as one-to-two-year business plans.
   iv. Long-term planning, such as the life of mine.
b. How are we going to represent these?
   i.   In a single orebody model.
   ii. In separate orebody models.
c. What is the operations ability to spatially and mechanically separate higher from lower grades and hence ensure a certain grade to mill? Known as SMU.
d. What grid sizes can be effectively estimated?
e. Is there sufficient data to regularise the data into larger-size blocks?
f. If regularised data is going to be utilised, do the regularised data reflect the true mean of the blocks, and if not, is it possible to Macro Krige to take cognisance of this?
g. If a single orebody estimation model is envisaged, how are we going to ensure that subsequent processes know and can correctly process the output data?

It is important to understand that there is no one grid size that provides a fit for all estimation (Annels, 1995). In the case of monthly planning and subsequent estimation of grade mined, one requires a high-resolution grid that does not necessarily extend much beyond the current face. High resolution implies that the grid size is small enough to effectively contain the required resolution of a smaller area/volume. Notwithstanding this, one needs to understand that there is an inverse relationship between resolution (how small a block is) and accuracy. This in turn means that not all smaller blocks can be utilised and those without the necessary accuracy be discarded. Hence, it is understood that the estimation grid of the smallest block size will not cover the entire lease area. For this to be accomplished, a larger block size covering a larger area will have to be estimated. However, this then begs the question as to whether the block satisfies certain criteria:

a. Is the estimated block size smaller than the SMU size?
b. If the answer to question (a) is in the affirmative? –

    i. At what stage does the estimates become smoothed to such an extent that they no longer represent a block smaller than the SMU?

    ii. Are you going to allow those estimates that have too low an accuracy due to smoothing to be? –

- Discarded and replaced by another block size.
- Determined and then smoothing dealt with on those blocks that have excessive smoothing.

c. If one is going to add an additional larger block size and this is going to be more than SMU size:

    i. What methods are you going to utilise to represent the estimates in a format indicating the volume and grade at SMU?

- A direct conditioning approach, which will require the smaller sized smoothed estimates and the larger sized blocks. Obviously, this in turn implies, that one has done the necessary work up front to determine what parameters and values of those parameters indicate a smoothed estimate. One does not want to unsmooth an estimate, which was not smoothed in the first place.
- A simple change of support indicating the volume and grade of the proportion of the large block size that will be available to mine.

d. Depending upon the sophistication of subsequent software utilised for mineral resource and mineral reserve estimation, do you require to add uneconomic ground that will unavoidably be mined? This in no way must be confused with selective mining considerations taken into consideration with the SMU. This is to satisfy the requisites of the code that requires that all sources of dilution be considered when estimating mineral resources and mineral reserves (Abzalov, 2013). Some of these may be obvious, such as the external waste that is mined when a minimum mining height is required which is greater in height than the height of the orebody. Other sources of dilution may not be as obvious and yet significant, such as the uneconomic or waste mined to get from one level to another via a raise, or the ledging required to install primary support or a slusher next to the raise. Although one may be aware of the grade, one cannot take this into consideration and hence is a source of dilution.

Only once all these criteria have been considered can one effectively kick off the estimation process via Kriging.

NUMBER OF SAMPLES REQUIRED TO OBTAIN A REPRESENTATIVE MEAN

This area of analysis has implications for many areas of mineral resource estimation and, as such, is an important part of the fundamental analysis required prior to more advanced estimation techniques. The areas that need to be covered are:

a. The number of samples required for a representative local mean to be used in Simple Kriging with a local mean.

b. The number of samples required and alternative methods to obtain a mean for a domain or zone.

c. The number of samples required for a representative mean of a regularised block. (This is important for semi-variograms from regularised data and if Makro Kriging is envisaged.)

CALCULATING THE REPRESENTATIVE LOCALISED MEAN

Here, we are attempting to determine the number of samples required for a localised mean to be representative of the area under question. Under these circumstances, it is important to understand that we are not attempting to obtain an estimate of the mean for the domain but a subset thereof. The idea of Simple Kriging with a localised mean is to attempt to reproduce a mean for Kriging that emulates localised conditions that may be produced due to anisotropy or other reasons, which may make the use of a global mean inappropriate. Considering Figure 6.1, it is observed that as the number of samples increases, the variance decreases up to a point and then increases once again. This occurs on numerous occasions until the variance finally levels off and does not rise again. At this stage, there are sufficient samples to obtain a representative mean for the area of interest. Prior to this, there is localised flattening of the variability, which equates to increasing area. In other words, the number of samples required to obtain a representative mean depends upon the size of the area in question. For example, larger areas require more samples for a representative mean. Although this fact is intuitive of nature very often little, or no consideration is taken of this fact.

How do we then obtain the minimum number of samples required for a representative mean? The answer depends upon the sample density one has on a particular operation. In the case of an operation that has a closely spaced sampling grid (e.g., 5 m × 5 m), an empirical methodology may be employed. Here, the data is regularised into a grid the size of the relevant search radius, the information for each block with regards to the number of samples within the block and the variance of the samples within each block needs to be available. An analysis is performed of the relationship between the 'in-block variance' as a function of the number of samples, and the number of samples where the variance plateaus is selected. In the case where there sampling is not sufficiently high in density, the procedure is slightly more complicated. For this case, a model of the point semi-variogram and a block of the dimensions of the required search radius are constructed. The in-block variance for various levels of block discretisation is subsequently computed (e.g. 1 × 1, 2 × 2, 5 × 5, 10 × 10, etc.) and analysed (Figure 6.2).

Typically, as the discretisation increases (equivalent to the number of samples within a block), so does the in-block variance. However, at some level of discretisation, the in-block variance plateaus. It is at this level that the addition of samples does not significantly change the value of the mean. This is the desired result. Consider the following four charts, which indicate the variance on the y-axis compared to the level of discretisation on the x-axis (Figure 6.2). Here, we are depicting four different increasing block sizes and the corresponding areas at which the variance plateaus. One will note that as the block size increases (from right to left and top to bottom), the level of discretisation required before the variance levels also increases from 6 to 10. In other words, at the smallest block size, one requires 6 × 6 = 36 samples for

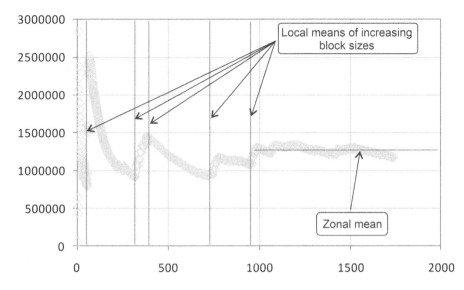

FIGURE 6.1 Graphical representation of the approximate local means versus estimation block sizes.

a representative mean, whereas at the largest block size one requires $10 \times 10 = 100$ samples to obtain a representative mean. However, one word of caution: this methodology assumes a regular grid of samples within the block size in question and does not account for sample clustering within a block. In the absence of a closely spaced grid of sampling it does, however, provide some level of rigour and analysis of the parameters required, which is most certainly far better than an uneducated guess.

OBTAINING A MEAN REPRESENTATIVE OF A DOMAIN OR ZONE

Follow the method discussed previously in the section covering declustering and determine whether the data has areas which are clustered; if so, determine the optimal declustering size and regularise data to the optimal declustered block size. Examine the number of regularised blocks and the distributional characteristics if there are sufficient blocks (approx. 100) and the distribution approximates normality, then the declustered mean is sufficiently accurate to be used as the zones mean. In the alternative case where there are insufficient numbers of blocks or the distribution is log-normal, use the existing *a priori* point variance, or if there are insufficient data to establish this, borrow the variance from a zone or domain that is considered sufficiently similar, and calculate a Sichel's mean estimate.

NUMBER OF SAMPLES REQUIRED FOR A REPRESENTATIVE MEAN OF A REGULARISED BLOCK

The method outlined in 'Calculating the representative localised mean' is used in this regard, with the area in question being the size of the block to which one is going to regularise. This information is then utilised to produce robust semi-variograms.

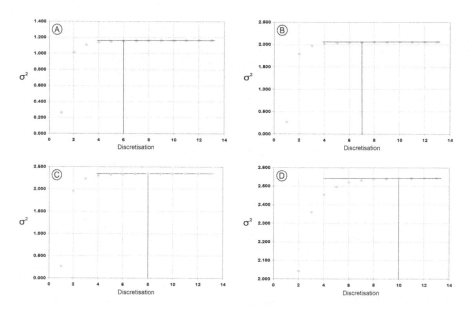

FIGURE 6.2 Illustration of optimum block discretisation selection versus in-block variance.

MACRO KRIGING

PROCESS FLOW MACRO KRIGING

a. Decide on the optimal block size for Macro Kriging, taking into consideration.
   i. Smaller block size. As we will have to subcell the Macro Kriging blocks into the size of the smaller blocks.
   ii. The optimal block size analysis.
b. Obtain the required number of samples for a representative mean.
c. Model the nugget effects using the block size obtained above and multiple instances of the number of samples.
   i. Calculate the point semi-variogram and then obtain the nugget effect of this.
   ii. Regularise data into the optimal block size using a minimum of two and a maximum of two samples.
   iii. Do the same as in *b* changing the minimum and maximum numbers for at least five different numbers of samples.
   iv. Calculate the semi-variograms for all the regularised files.
   v. Model the nugget effects for all the experimental semi-variograms calculated in point iv.
   vi. Plot the nugget effects on an Excel graph with the number of samples on the x-axis and the nugget effects on the y-axis. Remember that your point semi-variogram gets plotted on your graph as number of samples = 1.
   vii. Model the relationship between number of samples in a block and the nugget.

d. Regularise your file into the block size you want to use using no constraints on your maximum number of samples.
e. Calculate your experimental semi-variogram using the file you created in point (d).
f. Model your experimental semi-variogram calculated in point (e).
g. Using the regularised file created in point (d), create and populate a field in the file called macro nugget effect 'MKNUG' using the following logic:
    i. If the number of samples is less than that determined under point (b), then use the formula for the MKNUG field.
    ii. If the number of samples is greater than that determined under point (b), then use the nugget from your model you obtained in point (f).
h. From here onwards, proceed as per usual using Simple Macro Kriging.

CALCULATING THE NUGGET EFFECT FOR A MACRO KRIGING EXERCISE

As indicated previously, Macro Kriging takes into consideration the level of information of a block used within the Kriging environment. It accomplishes this by means of the fact that the nugget effect obtained from the semi-variogram for a certain-block size varies with the number of samples that exist within the blocks. Having said this, it then becomes an important part of Macro Kriging to understand how the nugget effect changes with the number of samples that are within a block. To obtain this relationship, the following methodology is followed.

Once having obtained the optimal block size for Kriging, as explained in the previous section, now:

a. Calculate and model the point semi-variogram for the domains.
b. Do multiple regularisations run with a minimum and a maximum number of samples per block set to 2, 5, 10, 15, etc., until finally, one reaches a stage where the number of samples is so high that insufficient number of blocks exist to model a semi-variogram, or the number of samples reaches that required for a representative mean of a block as outlined in the section under obtaining the number of samples for a representative mean. N.B. Remember to set the minimum and maximum number of samples to the same number to ensure that on a particular regularisation run, we have blocks that have only the stipulated number of samples within each block.
c. For each regularisation run results, calculate the experimental semi-variogram.
d. For each regularisation run model, the nugget effect of the semi-variogram.
e. Plot the results of the nugget effects on a graph and apply a model to the results (Figure 6.3). The equation from this model will be used to determine the nugget effect of each block for subsequent Macro Kriging.

MACRO KRIGING SEMI-VARIOGAMS AND MODELLING

All methods and checks for the calculation and modelling of semi-variograms outlined previously in the document still apply under these circumstances and full procedures must be followed.

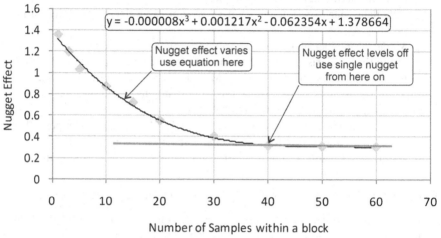

Number of Samples within a Block vs the Nugget Effect

$$y = -0.000008x^3 + 0.001217x^2 - 0.062354x + 1.378664$$

Nugget effect varies
use equation here

Nugget effect levels off
use single nugget
from here on

FIGURE 6.3   Number of samples within a block versus the nugget effect.

SUMMARY FOR MACRO KRIGING

Although Macro Kriging follows the same procedure as other Kriging methods, a certain amount of preparation before Kriging commences is required in these circumstances that do not occur normally. This relates to the variable nugget effect required on a block-by-block basis for the data to be used for the Kriging exercise. Here, we require the number of samples within each regularised block as a field in the regularised output file. The number of samples is then used as an input to the equation derived in the previous section, 'Calculating the nugget effect for a Macro Kriging exercise' to obtain the nugget effect for that block. This, in turn, is placed in a field called MKNUG and is used in the Kriging. Kriging is then run using the minimum and maximum sample criteria calculated in the section 'Future mining simulations or cross-validations'. The results are analysed, and any substitutions of Ordinary Kriging with Simple Kriging, as outlined in the same section, is applied. From this point onwards, the dispersion variance of the SMU-sized blocks within the Kriged size blocks can be calculated on a block-by-block estimate basis as outlined in the section calculating the dispersion variances. These dispersion variances can now be used in any subsequent grade-tonnage curve calculations for resources or theoretical reserves.

CONCLUSION

Geostatistical modelling, while widely used for estimating grade and tonnage of mineral resources and ore reserves, has its limitations when it comes to extrapolative predictions beyond the spatial bounds of the available data. This is because geostatistical methods rely on interpolative predictions based on inferred spatial variability of the existing data, such as variogram models. However, these techniques do not directly describe the physical processes governing the spatial distribution, making them less suitable for making predictions outside the range of the available data. Among the various spatial data interpolation techniques are nearest neighbour, IDW, radial basis functions, triangulation, and polygonal methods. In geostatistics, the most common estimation techniques used are Ordinary Kriging and Simple Kriging, which are variants of basic linear regression methods. These methods allow the estimation of a single regionalised variable in unsampled locations. One of the key features of Kriging is that it minimises estimation errors and ensures that the mathematical expectation of the estimation error is zero. Kriging assigns more weight to samples located closer to the estimated point and decreases the weights rapidly with increasing distance from the target point. Some samples may even receive negative weights due to a specific property of the Kriging system known as the screening effect. Negative weights can cause Kriging estimates to fall outside the range of the available data, offering the advantage of capturing potential extreme values beyond the sampled data. However, this can also lead to biased estimates, especially when dealing with high-grade samples.

To address the issue of negative weights and potential bias, various mathematical procedures can be applied. These include replacing negative weights with 0 and resetting positive weights to sum to 1, forcing the kriging weights to be positive by adding a constant equal to the modulus of the largest negative weight and resetting the weights to sum to 1, using statistical transformations, or imposing constraints directly on the Kriging estimates. Despite the advantages of Kriging, it is essential to consider the smoothing effect of this method. Kriging estimates typically lie between the largest and smallest values of the input data, resulting in a reduction in the variance of the estimated points compared to the data points. The degree of smoothing can vary depending on the search neighbourhood and variogram model used. Excessive smoothing of estimates can lead to inaccuracies, especially when estimating resources at a specific cutoff. This is because overly smoothed models may not accurately represent the spatial distribution of the attribute being estimated, particularly in cases where extreme values are of interest, such as the concentration of metallurgically deleterious components in an ore body or distribution of environmentally hazardous materials. While Kriging models lack global bias (the estimated global mean is accurate), they can still exhibit conditional bias, where the sign and value of the bias depend on the grade class. This is primarily caused by the smoothing effect of Kriging, which tends to overestimate low-grade values and underestimate high-grade values. It is worth noting that Kriging equations also underlie non-linear estimators and conditional simulation techniques, which can be explored in more detail in subsequent chapters.

REFERENCES

Abzalov, M. Z. (2013). Measuring and modelling of dry bulk rock density for mineral resource estimation. *Applied Earth Science, 122*(1), 16–29.

Abzalov, M. & Abzalov, M. (2016). Methodology of the mineral resource classification. *Applied Mining Geology, 12*, 355–363. https://doi.org/10.1007/978-3-319-39264-6.

Annels, A. E. (1995). The development of a Resource Reliability Rating (RRR) system for the evaluation of mineral deposits. In *Proceedings, mineral resource evaluation conference* (p. 16). Yorkshire: University of Leeds.

Annels, A. E. (1997). Errors and classification in Ore Reserve estimation: The Resource Reliability Rating (RRR). Proceedings, Assaying and Reporting Standards Conference. AIC Conferences, Paper no. 5, p. 29.

Annels, A. E., and Dominy, S. C. (2002). *Development of a Resource Reliability Rating (RRR) system for mineral deposit evaluation and classification.* In: Proceedings of Australian Institute of Mining and Metallurgy Conference 2002 (pp. 27–34). From: Value Tracking Symposium, 7-8 October 2002, Brisbane, QLD, Australia.

Deutsch, C. V. & Journel, A. G. (1998). *Geostatistical software library and user's guide* (2nd ed.). Oxford University Press, Applied Geostatistics Series, Deutsch.

Krige, D. G. (1996). A practical analysis of the effects of spatial structure and of data available and accesses, on conditional biases in ordinary kriging. In Baafi, E. Y. & Schofield, N. A. (Eds.), *Geostatistics Wollongong. Fifth international geostatistics congress* (pp. 799–810). Wollongong: Kluwer Press.

7 Post-mineral resources estimation processes

INTRODUCTION

In the discipline of mineral resource estimation, the absence of sufficient data in some regions of interest inevitably leads to estimates with heightened uncertainty. Ordinary Kriging, a widely used estimation technique, introduces a smoothing effect that can result in significant discrepancies between expected grades and true values (Abzalov, 2013). This discrepancy becomes particularly problematic as Kriging estimates fail to faithfully reproduce both the histogram and semi-variogram models of sample data. A meticulous post-estimation analysis is indispensable to address this challenge and ensure the reliability of estimates (Annels, 1995). Additionally, conducting follow-up sampling is highly recommended to refine further and validate the estimated values.

The integrity of the data forming the basis of the estimate is of paramount importance. The data must accurately represent spatial distribution and be truly representative of the entire population. Errors related to sampling, assaying, geological interpretations, and other sources can be introduced during the process of interpretation, estimation, and classification of mineral resources (Annels, 1997). Unfortunately, confidence limits are rarely provided in mineral reserve statements, and when they are, they often fail to adequately consider the multiple factors that contribute to uncertainty in grade and tonnage estimates. It is crucial to acknowledge that estimation inherently involves some level of risk and uncertainty (Dominy et al., 2002).

The accuracy of mineral resource estimation is fundamentally reliant on various factors, including the availability and quality of geological data, the exploration methods employed, and the effectiveness of sampling techniques. These aspects collectively contribute to determining the inferred, indicated, and measured categories of mineral resources, providing a basis for decision-making and project planning. Within this context, a thorough analysis of the selective mining unit (SMU) is essential to understand how it influences mineral resource estimation. The SMU's characteristics play a vital role in shaping resource estimates and significantly impact resource classification, conversion to mineral reserves, and overall project economics. Through focusing on the complexities of the SMU, mining professionals gain crucial insights into optimising mineral resource extraction strategies and making informed decisions.

Converting mineral resources to mineral reserves is a critical step in mining projects, and it involves a multidimensional estimation process. Factors such as mining and processing methods, metallurgical recovery rates, economic considerations, and environmental constraints all come into play (Deutsch and Journel, 1998).

DOI: 10.1201/9781032650388-7

143

A comprehensive understanding of these aspects ensures that mineral resources are accurately and responsibly classified as mineral reserves, forming the bedrock for profitable and sustainable mining operations. Dispersion variance, another important consideration, plays a significant role in mineral resource estimation. Mining professionals must carefully assess dispersion variance to manage risk effectively and make well-informed decisions. mineral resource extraction strategies can be optimised by comprehending the extent and impact of dispersion variance, leading to more efficient mining practices. As we focus more into post-mineral resource estimation, it becomes evident how accurate classification and an understanding of dispersion variance contribute to confidently converting mineral resources into mineral reserves. The success of this conversion process paves the way for sustainable and profitable mining ventures where responsible resource management is at the forefront.

CONSIDERATIONS FOR MINERAL RESOURCES CLASSIFICATION

In the realm of mineral resource classification, the primary objective lies in offering prospective investors and their advisors a comprehensive understanding of the reliability of mineral resource estimates. This confidence largely hinges on critical factors such as the geological model's trustworthiness, the consistency of mineralisation, the configuration of the sampling grid, the quality and quantity of sampling data, and the robustness of the estimation method. However, it has been observed that resource geologists often prioritise the confidence associated with grade estimates, often overlooking the underlying geological model's inherent risks (Dohm, 2004).

Over the years, regulatory advancements have consolidated various international and local reporting codes, ensuring clarity and coherence in mineral resource reporting on a global scale. The Committee for Mineral Reserves International Reporting Standards (CRIRSCO) has integrated minimum standards for Public Reporting of Exploration Results, Mineral Resources, and Mineral Reserves into the International Reporting Template. Additionally, it provides guidance and interpretive guidelines for the nations represented on the CRIRSCO committee. It is important to note that the South African Code for the Reporting of Exploration Results, Mineral Resources, and Mineral Reserves (SAMREC) aligns with CRIRSCO guidelines and is pertinent to South Africa, offering definitions for mineral resources and categorising them into Inferred, Measured, and Indicated categories based on the quality and confidence of geoscientific evidence. Below are some of the commonly used classification schemes:

a. Kriging slope of regression
    The phenomenon known as the regression effect has been acknowledged in Southern Africa for over six decades. Initially introduced by Krige (1951) to account for discrepancies between stope and development estimates of Witwatersrand gold grades, Till (1974) further developed the concept with the introduction of the reduced major axis regression. This refined regression approach aimed to achieve a better fit around the 45° line by utilising the standard deviations of the actual values and the estimates. The regression effect arises due to both the information effect, where incomplete information is available at the time of estimation, and the change of support

effect, stemming from block estimates being lower than sample values. As a result, the regression effect leads to overestimating high values and under-estimating low values, resulting in a regression slope of less than one. The impact of the regression effect diminishes when the slope approaches one, indicating greater confidence in the estimate (Dohm, 2016b). The regression slope offers insight into the extent of overestimation or underestimation in the evaluated blocks. Correcting for the regression effect relies heavily on the assumption that sample values follow a normal distribution, as variance and covariance hold limited meaning when applied to skewed data sets (Clark, 2015). Hence, data should be verified to follow a Gaussian distribution during exploratory data analysis before considering the regression slope for mineral resource classification.

b. Distance to nearest sample

Geometric techniques based on inter-sample spacing are very common in resource classification. One such method is using drill-hole spacing, which relies on the distance between boreholes surrounding the estimated block. This technique is straightforward when boreholes are vertical and evenly spaced (Silva and Boisvert, 2014). The method of classifying mineral resources based on the distance to the nearest sample has been employed by manganese mines since at least the 1970s (Beltrame et al., 1981). In this approach, the confidence level of the estimated block is influenced by its proximity to the nearest known sample. Sometimes, the average distance of all the samples is used instead. The block closest to the known sample is typically assigned the highest estimated confidence. However, it is important to recognise that this method has certain limitations, as it assumes continuity of mineralisation based solely on borehole spacing, which may not always be the case (Arik, 2002). Measured Resources are characterised by well-established global and local continuity, whereas Inferred Resources rely on apparent global continuity. Using borehole spacing as the sole criterion for mineral resource classification assumes three-dimensional continuity, which might not always hold true (Dominy et al., 2002; Silva and Boisvert, 2014).

c. Drill-hole spacing

One method employed in this realm involves classifying blocks by analysing the spacing between drill holes in the vicinity of the block under consideration. This technique is relatively simple to apply in straightforward cases with vertical and regularly spaced drill holes exhibiting minimal deviation. However, situations where drill holes are irregularly spaced, drilled in different directions, and possess significant deviations necessitate the calculation of drill-hole spacing locally within a search window. It is common for thresholds on drill-hole spacing to be determined based on historical experiences with analogous deposits at the discretion of the qualified person.

d. Search neighbourhood

Another approach involves classifying blocks based on distance and constraints pertaining to the number and configuration of data points within the search radius around the block. Typically, this technique involves defining estimation passes with varying search parameters. Blocks estimated by less restrictive passes are categorised as Inferred, while intermediate restrictive passes define the Indicated category, and the most restrictive pass identifies Measured blocks. Constraints considered in this method often include a minimum number of data points, drill holes, and informed octants. The qualified person makes informed decisions regarding appropriate thresholds based on their expertise.

e. Kriging variance

Kriging, an interpolation technique, aims to minimise the squared error between the estimated and true, unknown values. The resulting Kriging variance, representing the estimation error, depends solely on the estimation location, sample positions, and semi-variogram (Krige, 1997). Early Kriging estimates emerged from Matheron's work in 1971 (Deutsch et al., 2014), building upon the theories of Krige (1951) and Sichel (1973). Kriging offered not only block estimates but also a measure of estimate uncertainty in the form of the Kriging variance (Matheron, 1971). However, caution is warranted when employing Kriging variances for mineral resource classification, as Glacken (1996 cited by Snowden, 1996) warns that not all methods based on Kriging variances are valid. Arbitrary distribution percentiles should not be used for distinguishing mineral reserve categories, as emphasised by Arik (2002). Kriging variance is a widely used geostatistical practice for mineral resource classification. It represents the expected value of the squared error between actual and estimated grades, depending on block size, internal block discretisation, sample numbers, sample layout, and the semi-variogram. It offers relative confidence from block to block and helps define mineral resource classification categories (Snowden, 2001). However, for sparsely drilled deposits, using the distribution of Kriging variance alone can lead to erroneous classification as Measured for some parts of the mineral resource. It should be used with consideration of how well the drill-hole spacing addresses the deposit's geometry, spatial continuity, and overall data integrity. Special attention must be given to the relationships between Kriging variance, sample spacing, sample layout, semi-variogram ranges of influence, and mineral resource classification categories. Complex nested structures may significantly impact variability and should be prioritised when assigning confidence, with geological framework taking precedence over mathematical indicators (Snowden, 2001).

Emery et al. (2004) addressed concerns raised by Snowden (2001) and Glacken (1996) by applying Kriging variance as a mineral resource classification criterion for porphyry copper deposits. They integrated geometric and geological knowledge through geostatistical parameters to classify each block based on its Kriging variance. Ordinary Kriging is limited in recognising local data variability, particularly in heterogeneous deposits with low and high-grade domains (De Souza et al., 2004). The application of Ordinary Kriging variance as the sole method of mineral resource classification requires demonstrating homogeneity, considering the impact of contact metamorphism effects and faulting, which can introduce mineralogical heterogeneity. An inherent drawback of Kriging is its smoothing effect, particularly when dealing with non-Gaussian, highly skewed data.

Using Kriging on datasets with non-Gaussian distributions may result in non-reproducible spatial heterogeneity (Sadeghi et al., 2015). Shinozuka and Jan (1972) proposed Gaussian-based approaches. Several classification approaches rely on the definition of thresholds to differentiate categories. Using Kriging variance as a classification criterion offers the advantage of considering the spatial structure of the variable and the redundancy between samples. However, it can lead to undesirable artefacts, particularly near sample locations where the Kriging variance is very low, resulting in patches of Measured blocks in Indicated zones. Additionally, the Kriging variance does not account for the proportional effect, a common feature of earth sciences data, especially in high-grade zones where the variance tends to be higher.

f. Conditional simulation

Conditional simulation is preferable to Kriging for assessing the error in mineral resource estimates. Unlike Kriging variance, conditional simulation considers actual grades, making it more data-dependent and thereby considered an improvement. A method introduced by Dohm (2003) combines conditional indicator simulations for geology and conditional sequential Gaussian simulations for the grade, providing a means to assess the combined risk associated with geological interpretation and grade estimation. This technique considers the correlation between estimated blocks and accounts for the change of support, yielding valuable insights into the uncertainty of the estimate.

Conducting a substantial number of realisations from the same dataset is crucial in conditional simulation. These realisations are re-blocked into meaningful mining block sizes and shapes to represent expected block grades. The advantage of this approach lies in avoiding over-smoothing, ensuring that the simulated grades retain the same variability and continuity as the conditioning sample data. Additionally, it enhances the reliability of grade-tonnage curves for smaller blocks when compared to direct block estimates using available sample data (Dominy et al., 2002).

The growing support for conditional simulation as a mineral resource classification technique is attributed to its objectivity, enabling quantitative expression of the degree of risk. Categories based on conditional simulation incorporate grade variability and data location, a significant improvement over the Kriging variance-based approach (Snowden, 2001). However, it is essential to acknowledge that conditional simulation assumes normality and generalises the coefficient of variation, which might not hold for all distributions. Adequate assessments of distributions, including non-Gaussian ones, can be made with numerous realisations (Silva and Boisvert, 2014). There is a close relationship between mineral resource classification and production period, where shorter production periods entail higher confidence categories. Conditional simulation aids in determining expected coefficients of variation for estimated grades, tonnage, and metal contents during various production periods, thus enabling a quantifiable classification that recognises the uncertainty arising from subjective interpretations of available information (Dohm, 2004).

Wawruch and Betzhold (2005) have laid down widely accepted rules for the classification of base metals mineral resources. For measured resources, local variability is corrected to monthly or quarterly production units, with a confidence level of 90% and an estimation error within 15%. After the correction of local estimate variability to yearly production units, Indicated Resources are estimated within 15% error at a 90% confidence limit. Anything beyond these criteria is classified as Inferred (Deutsch and Deutsch, 2012; Wawruch and Betzhold, 2005). Conditional simulation garners support from several authors, including Wawruch and Betzhold (2005), Dohm (2005), Dominy et al. (2002), and Snowden (2001), as a superior approach to assess uncertainty compared to Kriging variance and other techniques. However, Deutsch and Deutsch (2012) recommend using it only as a supporting tool while maintaining geometric criteria for final classification, as classification results heavily rely on modeller assumptions and chosen parameters, potentially reducing transparency in resource disclosure to investors.

EXAMPLE OF CLASSIFYING KRIGED ESTIMATES

The need for a standardised and unbiased protocol for determining estimation confidence limits cannot be overemphasised. However, most geostatisticians will agree that the industry is far from establishing global criteria for the control of confidence limits and that those for Witwatersrand-type gold and mineral deposits in large, layered intrusions, such as the Bushveld Complex, may have to vary from those applied to a Petroleum Resource. It is still up to the industry to set up some form of quantifying the nature of our confidence in an estimate, and then translate these numbers into pre-defined confidence from Measured to Inferred. Increasingly of late, there appears to be a growing acceptance of methodologies that use the 95% lower confidence limits expressed as a percentage of the original estimate as a means of quantitatively indicating relative confidence; the confidence method utilises the lower confidence due to the fact that the risk is in obtaining a grade lower than that estimated. If one gets a higher grade than estimated here, one does not incur a financial loss if the estimated grade is above the cut-off.

This is not to say that there is not a potential risk of not considering a block just below the cut-off if there is a reasonably high probability of it being economic. It is this method that will be used in the following explanation. Although the authors are not prescribing the ranges to be utilised, the fact that one utilises a calculated confidence for each of the measured, indicated, and inferred categories ensures that there is consistency of method and effective comparisons can be made on an annualised basis. Additionally, any audit of generalised due diligence can be checked by an individual, and that person can achieve the same results given the same information. In the authors' opinion, this is a huge stride forward in achieving consistency in the field.

Here are example steps in calculating the confidence of Kriged estimates:

In order to calculate the confidence of an estimate, there are four pieces of information required; these are:

a. The estimate of the variable required.
b. The Kriging variance of the estimate in the space of the distribution of the variable. In other words, if the variable has a log-normal distribution, the variance required is the log variance.
c. A decision as to what confidence level one requires, for example, 90%, 95%, etc.
d. The inverse of the standard normal cumulative distribution function at the confidence level required. This will provide the number of standard deviations one has to go up or down from the estimate in order to obtain the required confidence.

The method utilised is as follows:

a. Calculate the input sample variance.
b. Test whether the input data distribution is Gaussian, uniform, log-normal, or 3-parameter log-normal.
c. If it is a 3-parameter log-normal distribution, calculate the optimal beta (third-parameter) and use it to transform the data into log space (e.g. LN(Au accumulation + Beta)).
d. Calculate transformed data mean and variance (e.g. log mean and log variance using transformation in step (c)).
e. Calculate the SMU Within Block Variance based on the point semi-variogram.
f. If the point-semi-variogram used for kriging points into blocks was normalised into 1, calculate the Between Block Variance (i.e. 1-Within Block Variance).
g. Calculate the log variance Between Blocks, which is done by multiplying the Between Block Variance obtained in step (f) with the log variance obtained in step (d).
h. Select the suitable confidence limit, for example, 95%, which is a z-score of 1.644853627.
i. For each estimated block, add a column named 'Log Kriging Variance' that calculates the multiplication of the Kriging variance with the log variance Between Blocks obtained in step (g).
j. Add a column to Calculate the Lower Confidence Limit of the estimates using the formula: =EXP((LN(Estimated grade value) – Log Kriging Variance/2) – confidence limit value of 1.644853627 × SQRT(Log Kriging Variance)).
k. Add a column for 'Percent Confidence' by dividing the Lower Confidence Limit from step (j) by the Estimated Ore Grade value and multiplying by 100.

Table 7.1 shows an example of the nature of output from classified Kriged output.

TABLE 7.1

Example of the nature of output from classified Kriged estimates

Block Number	X	Y	Z	Au value (cm.g/t)	Kriging Variance	LaGrange	Kriging Efficiency	Slope of Regression	Log Kriging Variance	LC CMGT	Percent Confidence	Total Sill or Sample Variance
0	5,800.90	−17,250.06	0.00	1,383.24	0.35	−0.34	0.65	0.74	0.09	818.33	59.16	429,183.61
1	5,800.90	−17,280.06	0.00	1,376.06	0.24	−0.24	0.76	0.81	0.06	893.11	64.90	Sample data Mean
2	5,800.90	−17,310.06	0.00	1,365.96	0.24	−0.24	0.76	0.81	0.06	890.80	65.21	541.98
3	5,800.90	−17,340.06	0.00	1,346.12	0.21	−0.22	0.79	0.82	0.05	902.26	67.03	Sample data Mean − Beta
4	5,800.90	−17,370.06	0.00	1,331.36	0.15	−0.19	0.85	0.84	0.04	956.46	71.84	502.07
5	5,800.90	−17,400.06	0.00	1,300.55	0.04	−0.14	0.96	0.89	0.01	1,105.56	85.01	Log variance of sampling
6	5,800.90	−17,430.06	0.00	1,310.97	0.13	−0.07	0.87	0.93	0.03	957.39	73.03	0.99
7	5,800.90	−17,460.06	0.00	1,553.45	0.20	−0.02	0.80	0.97	0.05	1,056.87	68.03	Beta
8	5,800.90	−17,490.06	0.00	1,341.82	0.14	−0.06	0.86	0.94	0.03	971.41	72.39	39.91
9	5,800.90	−17,520.06	0.00	583.31	0.09	−0.05	0.91	0.95	0.02	449.70	77.09	SMU within block variance
10	5,800.90	−17,550.06	0.00	333.03	0.06	−0.10	0.94	0.91	0.01	270.50	81.22	0.75
11	5,800.90	−17,580.06	0.00	486.50	0.06	−0.10	0.94	0.91	0.02	392.47	80.67	Between block variance
12	5,800.90	−17,610.06	0.00	609.07	0.00	−0.12	1.00	0.90	0.00	598.77	98.31	0.25

#												Log variance between blocks
13	5,800.90	−17,640.06	0.00	1,160.92	0.01	−0.11	0.99	0.91	0.00	1,055.08	90.88	0.25
14	5,800.90	−17,670.06	0.00	931.00	0.06	−0.15	0.94	0.88	0.01	761.51	81.79	Confidence limit
15	5,800.90	−17,700.06	0.00	615.26	0.12	−0.19	0.88	0.85	0.03	453.85	73.77	
16	5,800.90	−17,730.06	0.00	487.08	0.00	−0.10	1.00	0.92	0.00	459.60	94.36	95.00%
17	5,800.90	−17,760.06	0.00	610.70	0.06	−0.11	0.94	0.91	0.02	493.25	80.77	1.64
18	5,800.90	−17,790.06	0.00	924.09	0.07	−0.09	0.93	0.92	0.02	740.74	80.16	
19	5,800.90	−17,820.06	0.00	1,397.77	0.06	−0.09	0.94	0.92	0.02	1,129.83	80.83	
20	5,800.90	−17,850.06	0.00	1,217.51	0.05	−0.10	0.95	0.91	0.01	1,017.91	83.61	

TABLE 7.2
Example of confidence limit percentage ranges
for mineral resource classification

Confidence	Range (%)
Measured	65–100
Indicated 1	55–65
Indicated 2	45–55
Indicated 3	35–45
Inferred	Less than 35

Please ensure to follow the appropriate statistical and geostatistical principles while conducting these calculations and consider any specific requirements or standards relevant to the mineral resources classification process. Additionally, if any specific software or tools are to be used for the calculations, make sure to mention them in the protocol.

The estimation confidence limits will be expressed as per Table 7.2.

THE GEOLOGICAL CONFIDENCE

The geological confidence is obtained by taking into consideration the confidence associated with:

a. Continuity of the orebody
b. Structural confidence
c. Alteration index
d. Confidence in extrapolation or interpolation of facies types

In the case of points (a) and (b), these can be expressed by considering the level of detail available from existing drilling or, alternatively, the associated confidence extended to the survey if a 3D seismic survey has been completed. Point (c), in almost all cases, can be expressed by the level of drilling within the unmined areas. In cases where facies have been determined by indicator or multiple indicators, Kriging, as in Table 7.2, can be used.

OBTAINING THE OVERALL CONFIDENCE

The final confidence is obtained by combining the mineral resource estimation and geological confidence and accepting the lowest level as indicative of the confidence associated with that portion of the resource. This is accomplished with the following method:

a. Obtain estimation confidence via the method outlined previously within the document.
b. Obtain the geological confidence polygons from the Geology Department.
c. Do Indicator Kriging of facies into unknown areas superimpose these on top of the defined facies and then obtain the confidence on the facies boundary.

  d. Fill a model that uses the same prototype as the mineral resource estimation
     model with the geological confidences.
  e. Combine the continuity, structure, and facies confidences into a single model.
  f. Add the two models.
  g. Once the two models have been added together, run an algorithm that on a
     block-by-block basis accepts the lower confidence.

The above method will satisfy all the requisites required from the code.

A 'mineral resource' is a concentration or occurrence of solid material of
economic interest in or on the Earth's crust in such form, grade, quality, and
quantity that there are reasonable prospects for eventual economic extraction.
A mineral resource's location, quantity, grade, continuity, and other geological
characteristics are known, estimated, or interpreted from specific *geological
evidence and knowledge, including sampling.*

As can be seen, the three geological confidences and the mineral resource estimation confidence cover all the variables in bold italics in the definition.

SUB-ECONOMIC MINING FUNCTIONS

As discussed previously, not all lower-grade ore sent to the mill is covered by the
SMU selectivity. When considering the actual mining, one encounters areas in which
unavoidable mining below the economic break-even or cut-off that is not part of that
covered by the SMU needs to be mined. What we are modelling, in this case, is the
additional uneconomic material that is mined over and above that attributable to the
SMU. This comprises uneconomic material from the following type of sources:

  1. Raises and wide raises must be mined for the entire length from level to level
     to facilitate holing through ventilation and thus cannot be mined selectively.
  2. Raises that need to be ledged for mining infrastructure to be emplaced and
     thus cannot be mined selectively.
  3. Development that requires over-stoping to distress and as such, the eco-
     nomic value of the ground being mined cannot be taken cognisance of in
     the selection process.
  4. Areas in which the only access to economic ground is through mining the
     uneconomic area to reach the economic areas.

A further note to the above is that this relationship is modelled for use only in areas
where the resolution of the estimation is low, and a grade-tonnage curve approach is
required to determine the extent of the economic ground within a certain area. In the
case of high-resolution estimates the mine planner can determine where and what to
mine.

   The method employed is to regularise the data for the domain in question
into a block size equal to the SMU size, taking due care that the period used has

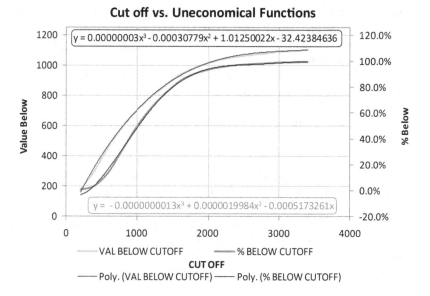

FIGURE 7.1   Approximation of the economically viable cut-off grade.

a similar economic break-even point to those being currently used. The logic behind this is that if no unnecessary uneconomic mining is occurring, then all the SMU-sized blocks mined would be pay. Then, rank the data values obtained from the regularisation from lowest to highest. One now counts the number of blocks below the economic break-even, and expresses these as a percentage of the total number of blocks; additionally, the average value of the blocks below the economic break-even is calculated. One does the process mentioned above for various cut-offs and plots the points on a graph. These can then be modelled using the standard trend modelling supplied by Microsoft Excel. See example in Figure 7.1.

Now that one has a model of the percentage uneconomic and uneconomic values for individual cut-offs, the proportions of uneconomic material that need to be included as being unavoidably mined can be calculated and included in Resource and Reserve calculations in areas where the resolution of the estimates militates against actual physical selection.

METHODS OF CHANGE OF SUPPORT

WHY DO WE NEED A CHANGE OF SUPPORT?

A change of support is required due to the volume–variance effect. In short, what this results in is that the larger the block size we consider, the larger the volume will be that one has to mine, and the lower the value will be that is effectively mined. However, if the block size considered exceeds the effective size at which one can mine, there will be a higher level of selectivity in actual production than indicated by the large blocks. This will result in one calling for a higher volume but lower value mined than what will be achieved. So, how do we calculate the actual value that one will be mining during the final production phase? This is accomplished by the change of support calculation.

A word of caution here – one must not merely apply a change of support and then think that one's work is now done. As intimated in the previous section, one then must apply the uneconomic mining that will be unavoidably mined as well.

WHAT PARAMETERS ARE REQUIRED FOR DOING THE CHANGE OF SUPPORT?

a. Estimated value of the large block.
b. Error in estimation of the large block (i.e. Kriging variance of the large block)
c. Error in estimation that occurs at the final stage before mining of an SMU-sized block (i.e. covered in section on estimation variance)
d. Variance of SMU-sized blocks in domain (i.e. obtained from Krige's Relationship)
e. Variance of large-sized blocks in domain (i.e. obtained from Krige's Relationship)

CALCULATING THE DISPERSION VARIANCES

Let us consider why we need to calculate the dispersion variances in the first place. This has got to do with the distribution of values within the orebody or any portion that may be considered an area of interest. As indicated in the previous section, it is important that we are aware of the actual level of mining selectivity that we can accomplish. This, in turn, will enable us to determine the level of resolution required in our estimation. However, in many cases, this level of resolution is not practically possible, and the estimated size of blocks far exceeds the required level. Nevertheless, it is required that we determine what it is that we are going to be able to mine within this larger area, given that our SMU is of a certain size.

Consider the sketch in Figure 7.2, where we have an area which is far more than the SMU size; this, in turn, has an average value of 1,321 cm.g/t. If we were to consider the SMU-sized blocks inside the big block, which are above a cut-off of, say, 1,500 cm.g/t, we would see that 28% of the area is mineable. In addition, the average grade of the mineable SMU-sized blocks inside the big block is 1,779 cm.g/t, thus meaning that we could mine the area in question with a margin of 19%. However, if we had simply looked at the average value of the large block, this was below the cut-off of 1,500 cm.g/t and hence we would not have decided to mine the area, resulting in a subsequent loss of revenue. If the calculations have been done with the required diligence, the result will be the proportion above the cut-off and the corresponding grade above cut-off. As the mathematics for these calculations are well established and extremely rigorous, in instances where the

Large Block Exceeding SMU Size

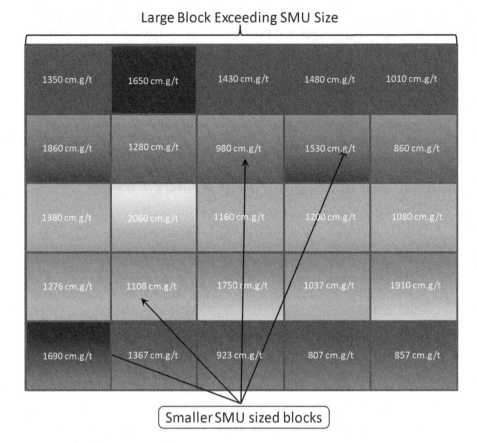

FIGURE 7.2 A sketch indicating economic and uneconomic SMU-sized areas within a large block.

corresponding mining does not bear out the results of the calculations, this is usually due to (see Figure 7.2):

a. An incorrect estimate of the mean of the large block.
b. An incorrect estimate of the variance of the SMU inside the block.
c. Not taking into consideration the information effect.
d. Using the incorrect SMU size. By far, this is the greatest contributing factor of all mentioned.

Here, we want to determine how the SMU-sized economic blocks are dispersed within the domain in general or in a certain block size that exceeds the SMU size. Under both sets of circumstances, we want to take into consideration the error that we will make in determining these blocks, hence the calculation of the information effect in the previous section. First, we will tackle the process of obtaining the variance of SMU-sized blocks within a domain. From Krige's Relationship, the dispersion variance of SMU-sized blocks can be obtained via the following formula. Dispersion Variance (DV) of SMU-sized blocks = Variance of points within domain which is the variance of points within SMU-sized blocks. However, this only takes cognisance of the dispersion variance where there is perfect information at the time of estimation, which is not the case in the real world. Here, we must take into consideration that our information available at the time of estimation will result in an error. The formula that takes cognisance of this is

$$DVF = VP - ViB - Ve \qquad (7.1)$$

where:

DVF = Final dispersion variance accounting for information effect.
VP = Variance of points within the domain.
ViB = Variance of points within the SMU-sized blocks.
Ve = Estimation variance of a SMU-sized block.
VP in this case, is merely the point variance of the samples within the domain.

ViB is obtained using the point semi-variogram model to obtain the in-block variance. Remember to use a large enough discretisation number to obtain the correct figure. Preferably graph various levels of discretisation as described in the section on obtaining a representative mean. Once again, it is important to note that if the values under consideration form a log-normal distribution, all calculations must be in log space. Second, having finished with the dispersion variance of the SMU within a domain, we will now deal with the dispersion variance of SMU-sized blocks within other blocks.

Let us consider the large blocks firstly. From Krige's Relationship, we can say:

$$\sigma^2 P \text{ in Domain} = \sigma^2 P \text{ in Large Blocks} + \sigma^2 \text{ between Large Blocks} \qquad (7.2)$$

Rearranging in terms of the variance of the points within the large blocks:

$$\sigma^2 P \text{ in Large Blocks} = \sigma^2 P \text{ in Domain} - \sigma^2 \text{ between Large Blocks} \qquad (7.3)$$

Now, considering the variance of SMU in large blocks:

$$\sigma^2 P \text{ in Large Blocks} = \sigma^2 P \text{ in SMU sized Blocks} + \sigma^2 \text{between SMU sized Blocks}$$

Now, substituting the right-hand side of Equation 7.3 into the left-hand side of Equation 7.4:

$$\sigma^2 P \text{ in Domain} + \sigma^2 \text{between Large Blocks} = \sigma^2 P \text{ in SMU sized Blocks}$$
$$+ \sigma^2 \text{ between SMU sized Blocks}$$

Rearranging equation in terms of SMU-sized blocks we now get:

$$\sigma^2 \text{ between SMU sized Blocks} = \sigma^2 P \text{ in Domain} - \sigma^2 \text{ between Large Block}$$
$$- \sigma^2 P \text{ in SMU sized Blocks} \qquad (7.6)$$

However, we need to remember to apply information effect thus the equation becomes:

$$\sigma^2 \text{ between SMU sized Blocks} = \sigma^2 P \text{ in Domain} - \sigma^2 \text{ between Large Block}$$
$$-\sigma^2 P \text{ in SMU sized Blocks} \qquad (7.7)$$
$$-\sigma^2 \text{ of the error on SMU size blocks}$$

Finally, one needs to take cognisance of the fact that an error is made on the estimation of the large blocks. Hence, the final equation for the dispersion variance of SMU-sized blocks within large blocks becomes:

$$\sigma^2 \text{ between SMU sized Blocks} = \sigma^2 P \text{ in Domain} - \sigma^2 \text{ between Large Block}$$
$$-\sigma^2 P \text{ in SMU sized Blocks} - \sigma^2 \text{ of the error on SMU size blocks}$$
$$+\sigma^2 \text{ of the error on estimation of the large blocks} \qquad (7.8)$$

Simplifying by substituting:

$$\sigma^2 \text{ between SMU sized Blocks in the domain in place of } \sigma^2 P \text{ in Domain}$$
$$-\sigma^2 P \text{ in SMU sized Blocks} \qquad (7.9)$$

Results in:

$$\sigma^2 \text{ between SMU sized Blocks in large blocks} = \sigma^2 \text{ between SMU sized Blocks in}$$
$$\text{the domain} - \sigma^2 \text{ between Large Block} - \sigma^2 \text{ of the error on SMU size blocks}$$
$$+\sigma^2 \text{ of the error on estimation of the large blocks}$$

$$\text{Eq. (7.10)}$$

Although, under most circumstances, this calculation will be accomplished behind the scenes, it is still important to understand the equation to check at least one or two of the calculations to ensure that the system being used is doing what we expect. These checks should be documented in order that subsequent audits may follow the same trail to validate the correctness of the system calculations (Figure 7.3).

CALCULATING THE CHANGE OF SUPPORT AT SMU

Change of support can now be accomplished by utilising the economic break-even or cut-off plus margin (see difference between economic break-even and cut-off in the section on these) and the dispersion variance in the change of support formula.

This will differ depending on whether the underlying distribution is log-normal or normal. Let us consider the standard Gaussian distribution calculation initially. Here, we calculate the percentage above an economic break-even or cut-off as follows:

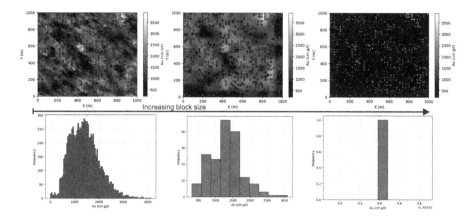

FIGURE 7.3 Illustration of the volume–variance relationship with increasing block size from left (ideal small block size) to right (larger blocks).

Percentage above the Economic break – even/cut off =

$$\text{normal cumulative distribution function of}\left(\left(\text{mean} - \text{cut off}\right)/\text{standard deviation}\right)$$

(7.11)

Whereas the equivalent value above cut off = Mean + standard deviation/
Percentage above Economic break – even or cut-off

$$\times \text{Normal probability distribution function of}\left(\left(\text{mean} - \text{cut off}\right)/\text{standard deviation}\right)$$

(7.12)

In the case of a 3-parameter log-normal distribution the calculation is slightly different.

First, one transforms the mean to a log mean by taking the Napierian logarithm of the untransformed mean plus the third parameter. Second, one then subtracts half the variance to obtain the transformed mean. The second step is because the back transform of a log mean is expressed as Untransformed mean = $e^{x+\text{logvarince}/2}$, hence, the subtraction of half the variance when going forward. The cut-off is also transformed into log space by taking the Napierian logarithm of the cut-off plus the third parameter.

One is now in the position to calculate the percentage above cut-off in the same way as with a normal distribution which is the normal cumulative distribution function of ((log mean – log cut-off)/log standard deviation)) (Equation 7.13).

The grade above cut-off is calculated as follows:

$$\text{Grade above cut-off} = \left(\text{Untransformed mean} + \text{third parameter}\right)$$
$$\times \text{normal cumulative distribution function of}$$
$$\left(\left(\left(\text{log cut-off grade} - \text{log mean}\right)/\text{log standard deviation}\right) - \text{log standard deviation}\right)$$
$$/\text{percentage above cut-off}) - \text{third parameter}$$

(7.14).

One now has an estimate of the proportion of economic ground and the grade thereof at the required SMU and cut-off.

CALCULATING THE ACTUAL RESOURCE AVAILABLE

Here too often the authors have observed the method of utilising the remaining ground *in situ* as being the resource. If one is considering an open pit example, this may suffice if the blocks going to the mill and low-grade stockpile are diligently annotated as such, and the software utilised takes cognisance thereof in the following manner. Ore with the prerequisite grade that is going to the mill and that includes all sources of dilution can immediately be classified as mineral resource at the confidence determined in sections first and second sections of this chapter. On the other hand, material going to the low-grade stockpile cannot be declared as a resource without further factors being considered. Here, one needs to determine an estimate of grade and volumes that will become available after the enrichment processes envisaged, such as chemical or biological leaching, which will leave a lesser proportion of the ore available to go to plant. Additionally, one needs to take cognisance of the variation in the leaching process and downgrade the confidence of this portion to ensure that this represents the overall level of confidence. Tonnage designated as waste can under no circumstances be classified as part of the resource with a single proviso, that being that it is part of an unavoidable dilution due to constraints imposed by mining or logistics. When considering an underground mine, here one is faced by a series of parameters that decide whether ore can be considered as part of the Resource:

1. Will the current owners of the company mine this portion of the orebody? What one has to be aware of and take into consideration, is that the Resource declared must provide an indication of the possible economic benefit to the current stakeholders and shareholders of the company. Hence, those areas that will not be mined by the current company cannot be declared as a Resource; this may become a resource for a future company that may buy the mine or portion thereof, and then it becomes their responsibility to declare it. This obviously leads one to the exception to this rule. Here, one may find an exploration company that has done sufficient exploration and drilling to estimate the values within the orebody; however, they may not be interested in or able to do the actual mining. Here, it is quite acceptable that the company in question obtain *a priori* information as to the production and capital costs involved and utilise these to provide an estimate of a potential Resource if it is to be mined by an alternative company. Due care must be taken to ensure that any costs involved are of a realistic nature and represent what most likely will happen if a mining company were to take over and mine. Nevertheless, one needs to understand that this Resource calculated is not a Real Resource but merely an estimate of such for future business decision-making.
2. Is it possible for the ground in question to be mined? Here, one may face a situation in which there has been extensive fall of ground which mitigates

against opening as the cost–benefit analysis determines that it would not be economically feasible. Alternatively, one may have a block of ground on a level that has been abandoned and stripped; once again, the cost of re-equipping may exceed the financial benefit or may detract from a more efficient spending of capital. Under these circumstances, the tract of ground in question must be marked as not available to mine and excluded from the Resource Inventory.

3. Those areas in which the estimates are of measured, indicated, and inferred confidence, but do not have the level of resolution to determine what will be mined and what will remain unmined, need the calculation of the proportion of SMU-sized blocks that will be mined as well as the additional dilution of the unavoidable uneconomic mining, as explained in the previous section. Here the tract of ground is annotated as such and only this proportion of ground is returned as part of the Resource.

4. There is a level of confidence that may occur within the lease area which, to all practical extents, is at a level of confidence equivalent to 'I have no clue'; these blocks must be annotated as such and not included as part of the Resource.

5. Tracts that are in the process of exploration but do not at present have the necessary levels of information are not to form part of the resource and must be clearly annotated as such. These block's tonnage can be reported upon, but it must be clearly stated that they are not included as part of the resource.

6. Losses of ground are not part of the Resource, and all due care must be taken to ensure that these are identified and indicated as such in as much detail as possible.

7. Areas of reef elimination, for example, waste on contact, cannot form part of the Resource and need to be clearly indicated as such so that they can be taken cognisance of and excluded.

8. Dip and strike stability pillars; here one is faced with the question as to whether the pillars in question will be mined at a later stage once ground stresses have abated and are lower at a later stage in mining. Nevertheless, these pillars need to be clearly identified and marked as either unavailable to mine (In which case they cannot be included as part of the Resource), or a Mineable Pillar which will be mined at a later stage. The need for this to be assigned to the individual pillar blocks is that, in the Reserve, this block confidence must be downgraded because all of the pillars may not be available at time of eventual mining. Hence a Proven Reserve that is a stability pillar will be reported as Probable.

9. Finally, the code stipulates.

A 'mineral resource' is a concentration or occurrence of solid material of economic interest in or on the Earth's crust in such form, grade, quality, and quantity that there are *reasonable prospects for eventual economic extraction*. The location, quantity, grade, continuity, and other geological characteristics of a mineral resource are known, estimated, or interpreted from specific geo-logical evidence and knowledge, including sampling. Here one needs to take into consideration the portion of the statement *reasonable prospects for eventual economic extraction* and as such, the ground selected to be reported as a Resource, together with all its sources of dilution, needs to be economic. If this is not the case, such a portion of standalone ground cannot be included as part of the Resource.

CONVERTING MINERAL RESOURCES TO MINERAL RESERVES

Here one is faced with two methods, the first being the one with the highest level of confidence, and the second method being that of obtaining the Reserve by calculation only utilising the attributes assigned to the individual blocks of the ground as indicated under the previous section on Resources.

In the first case, one does an actual mine plan, planning each piece of ground and scheduling it according to the period in which it becomes available to mine.

Dip and strike stability pillars as well as bracket pillars are excluded from the schedule according to latest geological mapping and interpretation of structure as well as adhering to rock engineering principles and guidelines.

This then becomes known as the Scheduled Reserve. Once the tonnages and grades have been obtained of the areas scheduled, the additional Reserve parameters are then applied; these follow the ore flow as explained in the section under reconciliation. Here, the historical (not the planned) parameters such as mine call factor, tonnage discrepancy and plant recovery percentage are applied in the sequence encountered in the ore flow, producing the scheduled reserve from the Estimated Resource. Additionally, the Reserve confidence (except in those exceptional cases mentioned previously) is obtained from the Resource Confidence, with measured confidence translating into a Proved Reserve and an indicated confidence translating into a Probable Reserve. Inferred Resources cannot and should not be reported as Reserves. Once the areas under question have sufficient confidence due to additional sampling, drilling, etc., the Reserve can be reconciled as being, for example, Probable Reserves resulting from Resource upgrade from Inferred to Indicated.

CONCLUSION

In this chapter, we demonstrated that the post-mineral resources classification processes play a crucial role in refining and enhancing the accuracy, reliability, and usability of mineral resource models. Through the application of various post-processing techniques, mining companies can gain deeper insights into their mineral deposits, leading to informed decision-making, optimal resource management, and the conversion of mineral resources into economically viable mineral reserves. The utilisation of model transfer functions emerges as a powerful tool, enabling volumetric calculations and validating ore grade. These transfer functions facilitate critical decision-making processes and provide valuable inputs for mine planning and production strategies. Central to the post-processing stage is the classification of mineral resources based on their geological confidence. This classification considers the continuity of the orebody, the structural confidence, and the assessment of alteration indices. By accurately categorising mineral resources as measured, indicated, or inferred, mining companies can effectively gauge the level of confidence and reliability associated with each resource estimate. A prudent consideration of the confidence in interpolation of mineral resources classification method is the accuracy to represent uncertainties in the estimates using objective quantifications and semi-quantitative categories to distinguish between measured, indicated and inferred mineral resources. This step acknowledges the importance of extending geological characteristics beyond the available data points and evaluates the robustness of the resource estimation in areas with limited data. The ultimate objective of post-mineral resources classification is to transform mineral resources into mineral reserves. This important step involves a comprehensive assessment of technical, economic, and environmental factors, ensuring the feasibility and profitability of extracting the mineral resources. Compliance with regulatory requirements is also paramount to responsibly manage the extraction process. Collaboration between geologists, geostatisticians, and other experts is fundamental throughout the post-processing phase. The fusion of geological knowledge, statistical analysis, and mining considerations culminates in robust and comprehensive resource models. This collaborative effort ensures that mining companies can make well-informed decisions, optimise their resource management strategies, and sustainably exploit their mineral deposits for the benefit of all stakeholders. We therefore emphasise that the post-mineral resources classification processes are an indispensable component of the mining value chain. They empower mining companies to unlock the true potential of their mineral resources, identify and mitigate risks, and make strategic decisions that drive the successful and responsible operation of mining projects.

REFERENCES

Abzalov, M. Z. (2013). Measuring and modelling of dry bulk rock density for mineral resource estimation. *Applied Earth Science, 122*(1), 16–29.

Annels, A. E. (1995). The development of a Resource Reliability Rating (RRR) system for the evaluation of mineral deposits. In *Proceedings, Mineral Resource evaluation conference* (p. 16). Yorkshire: University of Leeds.

Annels, A. E. (1997). Errors and classification in Ore Reserve estimation: The Resource Reliability Rating (RRR). In *Proceedings, assaying and reporting standards conference* (p. 29). AIC Conferences, Paper no. 5.

Arik, A. (2002). Comparison of resource classification methodologies with a new approach. 30th International Symposium on the Application of Computers and Operations Research in the Mineral Industry. Symposium Proceedings, 2002. Phoenix, Arizona. APCOM, pp. 57–64.

Beltrame, R. J., Holtzman, R. C., & Wahl, T. E. (1981). *Manganese resources of the Cuyuna Range, East-Central Minnesota*. Minnesota: University of Minnesota.

Clark, I. (2015). Regression revisited (again). *Journal of the Southern African Institute of Mining and Metallurgy, 115*(1), 45–50. ISSN 2411-9717.

Deutsch, J. L. & Deutsch, C. V. (2012). A new version of kt3d with test cases. *Centre for Computational Geostatistics Annual Report, 14*, 403–410.

Deutsch, C. V. & Journel, A. G. (1998). *Geostatistical software library and user's guide* (2nd ed.). Deutsch: Oxford University Press, Applied Geostatistics Series.

Deutsch, J. L., Szymanski, J., & Deutsch, C. V. (2014). Checks and measures of performance for kriging estimates. *Journal of the Southern African Institute of Mining and Metallurgy, 114*(3), 223–223.

De Souza, L. E., Costa, J. F. C. L., & Koppe, J. C. (2004). Uncertainty estimate in resource assessment: A geostatistical contribution. *Natural Resources Research, 13*(1), 1–2.

Dohm, C. E. (2003). Application of simulation techniques for combined risk assessment of both geological and grade model – An example. *Journal of the Southern African Institute of Mining and Metallurgy, 155*, 351–354.

Dohm, C. E. (2004). Quantifiable Mineral Resources classification: A logical approach. *Quantitative Geology and Geostatistics, 14*(1), 333–342.

Dohm, C. (2005). Quantifiable Mineral Resource Classification: A Logical Approach. In: Leuangthong, O., Deutsch, C.V. (eds) Geostatistics Banff 2004. Quantitative Geology and Geostatistics, vol 14. Springer, Dordrecht. https://doi.org/10.1007/978-1-4020-3610-1_34

Dohm, C. E. (2016b). *MINN7006: Geostatistical methods in Mineral Resource evaluation. [Class notes 05–07].* University of the Witwatersrand.

Dominy, S. C., Noppé, M. A., & Annels, A. E. (2002). Errors and uncertainty in Mineral Resources and Ore Reserve estimation: The importance of getting it right. *Exploration Mining Geology, 11*(1–4), 77–98.

Emery, X., Ortiz, J. M., & Rodriguez, J. J. (2004). *Quantifying uncertainty in Mineral Resources with classification schemes and conditional simulations.* Chile: University of Chile.

Glacken, I. (1996). Change of support by direct conditional block simulation. M.Sc. Thesis, Stanford University.

Krige, D. G. (1951). A statistical approach to some mine evaluation and allied problems on the Witwatersrand. M.Sc. Thesis, University of the Witwatersrand.

Krige, D. G. (1997). Block Kriging and the fallacy of endeavoring to reduce or eliminate smoothing. Keynote address, 2nd Regional APCOM Symposium, Moscow State Mining University, August 1997.

Matheron, G. (1971). *The theory of regionalized variables and its applications.* Paris: École Nationale Supérieure des Mines de Paris.

Sadeghi, B., Madani, N., & Carranza, E. J. M. (2015). Combination of geostatistical simulation and fractal modelling for Mineral Resource classification. *Journal of Geochemical Exploration, 149*, 59–73.

Shinozuka, M. & Jan, C. M. (1972). Digital simulation of random processes and its applications. *Journal of Sound and Vibration, 25*, 111–128.

Sichel, H. S. (1973). Statistical valuation of diamondiferous deposits. *Journal of South African Institute of Mining and Metallurgy, 73*, 235–243.

Silva, D. S. F. & Boisvert, J. B. (2014). Mineral Resource classification: A comparison of new and existing techniques. *Journal of the Southern African Institute of Mining and Metallurgy, 11*, 265–273.

Snowden, D. V. (1996). Practical interpretation of resource classification guidelines. In *AusIMM Annual Conference "Diversity, the key to Prosperity".* Perth, 1996. https://citeseerx.ist.psu.edu/document?repid=rep1&type=pdf&doi=7affe2f0b5ec390abf4fd8fa7778426937a9d2be

Snowden, D. V. (2001). Practical interpretation of Mineral Resource and Ore Reserve classification guidelines. Mineral Resource and Ore Reserve Estimation. The Australian Institute of Mining and Metallurgy (AusIMM) Guide to Good Practice. Melbourne, pp. 643–652.

Till, R. (1974). *Statistical methods for the Earth scientist*. Palgrave, MacMillan.

Wawruch, T. M. & Betzhold, J. F. (2005). Mineral Resource classification through conditional simulation. In Leuangthong, O. & Deutsch, C. V. (Eds.), *Quantifiable Mineral Resource classification: A logical approach. Quantitative Geology and Geostatistics* (Vol. 14). Springer, Dordrecht. https://doi.org/10.1007/978-1-4020-3610-1_48

8 Nonlinear geostatistical methods and conversion of mineral reserves

INTRODUCTION

In mining geostatistics, Linear and Log-normal Kriging were the first methods used to estimate *in-situ* Mineral Resources in blocks or panels. Nonlinear methods and Indicator Kriging approaches were subsequently developed to measure the portion of a block that can be recovered when a cut-off is applied to selective mining units (SMUs). In practice, these methods are often based on Gaussian models with log-normal or normal score transformations or on indicators above cut-off grade.

Jacques Rivoirard

In the ever-evolving field of mining and mineral resource estimation, accurate assessments of mineral reserves play a critical role in decision-making and economic viability. As orebodies become increasingly complex and challenging to characterise, relying on traditional linear geostatistical methods has shown limitations. Enter the realm of nonlinear geostatistical methods which are cutting-edge approaches that promise to revolutionise the way we analyse and convert mineral reserves. This chapter covers the principles and applications of nonlinear geostatistical methods, exploring their advantages over linear methods, and the unique insights they provide. The journey begins with an elucidation of the shortcomings of traditional linear techniques, and the emerging need for more flexible, adaptive models to capture the intricate spatial patterns inherent in mineral deposits.

We then embark on a comprehensive exploration of the various nonlinear geostatistical methods at our disposal. From Indicator Kriging and multiple-point statistics to object-based and geostatistical simulation techniques, each method offers distinct advantages in tackling different geological scenarios. Through detailed case studies and real-world examples, we illustrate how these nonlinear approaches capture complex spatial relationships and reduce uncertainty in mineral resource estimation. Furthermore, this chapter delves into the critical process of converting mineral resources into reserves, which is an important step in mining projects that demands precision and prudence. We examine how nonlinear geostatistical methods not only enhance the estimation of reserves but also facilitate risk analysis, aiding mining companies and investors in making informed decisions with greater confidence. Our exploration of this topic would be incomplete without addressing the challenges and

DOI: 10.1201/9781032650388-8

potential pitfalls of implementing nonlinear geostatistical methods. We shine a light on data requirements, computational complexities, and best practices to ensure reliable and robust outcomes, minimising potential errors in the estimation and conversion process.

SEQUENTIAL GAUSSIAN SIMULATION

Sequential Gaussian simulation is a stochastic simulation method that uses the Kriging mean and variance to generate a Gaussian field, which is used to model the observed data (Deutsch and Journel, 1992). It is similar to geophysical inverse modelling in that a realisation is constructed satisfying some notion of similarity with actual observation. In this sense, sequential Gaussian simulation constructs stochastic realisations of a property that preserves the property's Kriging mean and variance and are therefore equiprobable given the observation. It is mostly used for geostatistical simulation of hydrocarbon reservoirs for continuous properties such as porosity in lieu of traditional deterministic interpolation-based approaches such as Kriging. Similar to other inverse modelling simulations, two key ideas exploited by sequential Gaussian simulation are that there are many solutions that satisfy a given set of constraints (of realism) that can yield the observed samples, and that it is possible to sample probable solutions using stochastic modelling. In our case, sequential Gaussian simulation allows us to examine different possibilities, especially resource heterogeneities that are not well-sampled by the physical data and provide a notion of uncertainty in the resource models by the standard deviation of all simulated values at any location.

From a mathematical perspective, sequential Gaussian simulation provides a mechanism to overcome the property of spatial interpolation techniques such as Kriging to yield artificially smooth resource models, which are physically unrealistic. The average of a number of realisations can then be interpolated to provide a resource model or map. When the number of simulations is very large and approaches infinity, the average of the realisations converges to the Kriged solution. The ability of sequential Gaussian simulation to generate additional equiprobable realisations makes it especially useful to emulate the dispersion variance at a level which is like that obtained under well-informed data situations (Krige et al., 2004), which is the variation of the resource at the block level. For mineral resource modelling, knowing this dispersion variance means that a global grade-tonnage curve can be modelled. Simulated values are Kriged estimates plus a stochastic component that corrects for smoothing in data and missing variance (Deutsch and Journel, 1992).

Data pre-processing is necessary for sequential Gaussian simulation and since Kriging would be used to obtain a final representation, the data requirement is similar to that of interpolation-based modelling and include outlier treatment and debiasing. The remaining steps to perform sequential Gaussian simulation are presented by Deutsch and Journel (1992) and a workflow diagram to illustrate the process is given in Figure 8.1 and include:

a. Calculate histogram and statistical parameters of raw data. This helps in understanding the characteristics of data, such as grade distribution.

b. Transform data into Gaussian space. This helps to preserve the global distribution of data.

c. Calculate and model variogram using the normal transformed data.

d. Define a grid.

e. Choose a random path. From here onwards, the process is that of the algorithm that is used for the sequential Gaussian simulation.

f. Krige a value at each node from all other values (known and simulated) to get mean $Z(x_P) = w_i Z(x_i)$ and its corresponding Kriging variance, $\sigma_K^2(x_p)$.

g. Draw a random value from a Gaussian distribution with a mean of 0 and variance of $\sigma_K^2(x_p)$, known as the residual $R(x_P)$. This retains the dispersion variance.

h. Add the Kriged estimate and residual to get simulated value $Z_{\text{sim}}(x_P) = Z(x_P) + R(x_P)$.

i. The simulated value $Z_{\text{sim}}(x_P)$ is added to the current dataset to ensure that the covariance and predictions are based on it. This makes the simulation sequential by making use of previously simulated values as subsequent input.

j. Randomly visit other nodes sequentially and simulate values at each node to avoid artefacts of deterministic or correlated search. A random sequence ensures that no artificial structures are introduced in the data and that each simulation has a significant difference. Previously simulated nodes are used as input data values during Kriging; hence the proper covariance structure between the simulated values is preserved.

k. Back transform all simulated values (in this step, a realisation is generated).

l. Repeat steps (a) to (i) to generate as many realisations as needed.

Validation of sequential Gaussian simulation results is necessary to identify whether sufficient realisation have been achieved. There are several methods to validate simulation results and a combination of these methods adds confidence to any sequential Gaussian simulation.

The validation methods include but are not limited to:

a. Comparing ergodic/non-ergodic variograms with declustered normal score transformation semi-variogram and cumulative distribution functions. Ergodicity is a statistical concept formulated by Ludwig Eduard Boltzmann, an Austrian physicist, and the founder of statistical physics. In geostatistics, if the experimental semi-variograms of all realisations converge to that of the original data, the sequential Gaussian simulation is ergodic. Otherwise, the simulation is non-ergodic. Non-ergodic simulations cannot realise all possible system configurations and are undesirable.

b. Comparing the mean and variance of the total realisations with the corresponding mean and variance of the block data;

c. Directions of anisotropy should be evident when examining any individual plot of a single realisation.

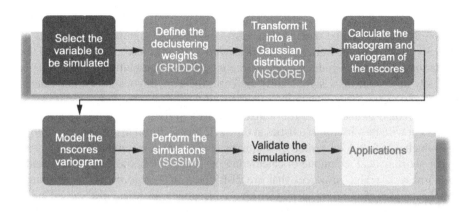

FIGURE 8.1   Sequential Gaussian simulation workflow.

d. The experimental semi-variogram calculated from the realisation results should provide a model that mirrors the model utilised in the simulation run, and

e. Constructing post-simulation models such as E-type models (i.e., local conditional expectation) and conditional variance. This is done by either averaging all blocks, which is equivalent to simple or ordinary values or use individual blocks to check whether errors on each simulation differ. If the uncertainty is equiprobable, the average errors will be the same throughout the grid realisations.

What one must understand with sequential Gaussian simulation in practice is that no single realisation expresses a best estimate. Rather, it presents a potential picture of the orebody if certain errors were realised. An area where the results of a sequential Gaussian simulation can be of immense use, is if one has a mine plan and one plans to do a Monte Carlo simulation in order to determine the project risk. Here one will evaluate the mine design and schedule multiple times utilising the different realisations and save the results thereof in the Monte Carlo algorithm. One can then randomly select the results of an individual realisation as well as randomly selecting the other Monte Carlo parameters thus producing a series of output NPVs that provide an estimate of the probability of achieving a positive NPV or a NPV that meets a certain criterion while at the same time taking orebody risk into consideration.

CALCULATION OF RECOVERABLE RESERVES

After determining a suitable model through either deterministic or stochastic approaches, it is important to estimate a cut-off grade to determine the quantity of Recoverable Reserves (Remacre, 1987). This calculation should consider the effects of the mismatch between the actual quantisation of material removal (e.g. a selective mining unit or SMU) and the resource model spatial quantisation. The SMU concept should be considered in assessing mineral resources, if the extraction quantisation differs substantially from the quantisation in the resource model, since statistical characteristics of resources and particularly the recoverable reserve change depending on the size of spatial quantisation. This could lead to significant deviations of critical operational constraints that are derived from the resource model, such as the grade-tonnage curve. Producing a resource model at the resolution of the SMU is typically unfeasible, as it significantly increases the sampling resolution and, therefore, the budget associated with the sampling strategy.

One method to ensure that the statistics derived at the resource model level are consistent with the observed statistics at the SMU level is through uniform conditioning, which was first documented by Matheron (1974). Uniform conditioning is a non-linear geostatistical technique of calculating the tonnage (Tv) and the mean grade (Mv) of recoverable resources that are distributed in a large panel (V) through small partitions of size (v), each of which is conceptually identical to a SMU (Chiles and Delfiner, 1999; Rivoirard, 1994; Wackernagel, 2003). Uniform conditioning uses the 'conditionally unbiased' large-block estimator to condition the average of a distribution of small blocks, thereby maintaining the correct grade-tonnage curves and applying a conditional distribution to obtain an accurate distribution of SMU grades (Assibey-Bonsu, 1998; Neufeld et al., 2005).

In a simplified form, uniform conditioning helps to determine what one will find when one gets to a particular mining area without the finer spatial grade resolution. Uniform conditioning estimates the distribution of grades and tonnages at an SMU above a selected cut-off (see section on cut-off grade) for a large size. As mentioned, it does not provide individual grades at an SMU size; rather, it provides an indication of how much one should expect from the block when the resolution is obtained, and planning can be achieved at an SMU size. This is achieved by calculating the grade-tonnage curve for each large-sized block based upon the value of that block and the calculated dispersion variance of the SMU-sized blocks within the large blocks. One then selects the appropriate cut-off from the grade-tonnage curve and assigns the value and proportion to the block (see the section on grade-tonnage curves).

Each panel $V \in \{X_p\}$ contains a total resource $Z(V)$, which contains a number of smaller blocks v, each of which contains a resource of $Z(v)$ that are classified as 'ore' if $Z(v)$ is greater than or equals to a cut-off value zc, and otherwise, it is classified as 'waste'. Uniform conditioning requires a number of steps (Figure 8.2): estimation of panel grades $Z(V)$, through either deterministic or stochastic methods; data transformation of the panel grades through Gaussian anamorphosis (Rivoirard, 1994; Wackernagel, 2003); change of support from point data to block data; estimate new cut-off grades and $Z(v)$; and determine the proportion and quantity of resource above cut-off (Abzalov, 2006). The Gaussian anamorphosis uses Hermite polynomials (H_k), which are built from the derivatives of the Gaussian density function and therefore Hermite polynomials can be conveniently derived using a recurrence relation. Using an Nth order Hermite polynomial and coefficients φ_k, $Z(v)$ can be transformed into the Gaussian variable $Y(v)$ by

$$Z(v) = \Phi\big(Y(v)\big) \approx \sum_{k=1}^{N} \varphi_k H_k\left(Y(v)\right) \qquad (8.1)$$

Thereafter, the change of support can be incorporated into the Gaussian model by a parameter r and therefore, with the change of support.

$$Z(v) = \Phi(Y(v)) \approx \sum_{k=1}^{N} r^k \varphi_k H_k \left(Y(v)\right) \tag{8.2}$$

Similarly, for panel grades and a panel change of support coefficient s,

$$Z(V) = \Phi(Y(V)) \approx \sum_{k=1}^{N} s^k \varphi_k H_k \left(Y(V)\right) \tag{8.3}$$

If the estimation of the resource model and, therefore, the cut-off grade occurred in original grade units, then it is necessary to transform the cut-off grade using Equation 8.3, e.g., $yc = \phi^{-1}(zc)$. The tonnage above cut-off is given by:

$$Tv(zc) = E\left[I_{Z(v)\geq zc} \mid Z(V)\right] = E[I_{Y(v)\geq yc} \mid Y(V)] = 1 - G\left\{ \frac{yc - (s/r)Y(V)}{\left(1 - (s/r)^2\right)^{\frac{1}{2}}} \right\} \tag{8.4}$$

The quantity of metal extractable can be given by:

$$Qv(zc) = E[Z(v)I_{Z(v)\geq zc} \mid Z(V)] = \sum_{k=1}^{N} \left(\frac{s}{r}\right)^k H_k \left(Y(V)\right) \sum_{j=1}^{N} \varphi_j r^j \int_{yc}^{\infty} H_k(y) H_j(y) g(y) dy$$

$$\tag{8.5}$$

Therefore, the mean extractable grade is $Mv(zc) = Qv(zc) / Tv(zc)$. Additional correction for insufficient information, for, regarding the dispersion variance at the resource block-model level, known as information effect, is usually applied to further improve a given change-of-support model (Journel and Kyriakidis, 2004; Wackernagel, 2003).

The performance of uniform conditioning may be assessed by how closely the conditioned grade-tonnage estimate conforms to the sequential Gaussian simulation model(s) and the Ordinary Kriging model. Table 8.1 shows the advantages and disadvantages of linear versus nonlinear geostatistical models.

Another method for calculating recoverable resources is localised uniform conditioning. Localised uniform conditioning was developed to determine the grades of small blocks when one was faced with an area of high data paucity and Kriging estimates would be severely smoothed (Abzalov, 2006). The method is a modified version of uniform conditioning which calculates the grade distribution for the large blocks and then separates these into small blocks via Kriging of the small SMU-sized blocks directly and then sorting these in increasing grade for each big block. From these sorted blocks, a grade can be calculated for each block within the large block

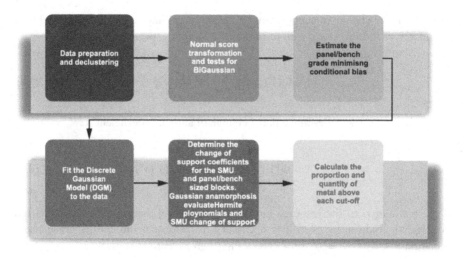

FIGURE 8.2   Uniform conditioning workflow.

from the uniform conditioning model of the grade-tonnage curve of each large block. Some of the pros and cons of this method are as follows:

a. Advantages
   - One gets grade-tonnage curves that represent the dispersion variance of the small blocks in question.
   - One achieves estimates of the SMU size block grades showing areas available for mine planning.
b. Disadvantages
   - The grades obtained for SMU-sized blocks are dependent upon the ranking of the values. If these values are correctly ranked one achieves correct estimates; however, if the ranking achieved is not correct, one achieves incorrect estimates.
   - The dispersion of the SMU-sized blocks and hence the estimates of the values of the blocks is dependent upon one having a decent estimate of the true dispersion variance, which may not be the case if one does not have sufficient data for this.
   - One is utilising a uniform dispersion variance of the blocks in question. The method of localised direct conditioning was devised to counteract this potential drawback (Krige et al., 2004).

CONVERTING RESOURCES TO RESERVES

After the estimation of mineral resources, the next step is to convert these resources to reserves. This conversion is necessary for reporting purposes and to enable economic assessment of a mining project. The Committee for Mineral Reserves International Reporting Standards (CRIRSCO) provides guidelines for the conversion of Mineral Resources to Mineral Reserves. In this chapter, we will discuss the CRIRSCO guidelines for converting Resources to Reserves.

TABLE 8.1
Assessment of advantages and disadvantages of the different mineral resource estimation algorithms

Characteristics	Ordinary Kriging	Sequential Gaussian Simulation	Uniform Conditioning
Advantages	Minimises the interpolation error and error variance. Is deterministic, only a single realisation is required. Provides uncertainty of the estimate in a form of Kriging variance. Compensates for the effects of data clustering. Provides the best local estimate given the available data. Provides an accurate global uncertainty including in areas that are poorly sampled.	Provides realistic grade-tonnage calculations. Provides alternative realisations and characterise heterogeneities. Honours spatial variability of global dataset. Provides an accurate global uncertainty including in areas that are poorly sampled. Appropriate for data with highly contrasting values since it reproduces the extreme values as well as their variability.	Models capture the variability of a mineral deposit at the mining site. Models capture the variability of grade at SMU sizes, e.g., a change of support. Uniformly conditioned blocks represent what can be economically recovered. Provides grade-tonnage curves at different SMU sizes to assess the recoverable resources at different scales.
Disadvantages	Localised smoothing, e.g., it does not produce estimates that represent values matching or exceeding the highest value of the deposit and does not reproduce zero grades. Unsuitable to directly estimate targets without data transformation and processing and where the extreme values are important. However, this can be mitigated using Indicator Kriging where one is interested in extreme values. Spatial quantisation often mismatches with the SMU size. Requires additional methods to quantify recoverable resources.	High estimation error when only a few realisations are used. High variability between simulations if conditioning data used is sparse. Needs caution in interpreting confidence limits calculated from post-processing simulation since uncertainty in the conditioning data can sometimes be large. Selection of small search neighbourhood can lead to poor conditioning and non-replication of the semi-variogram.	Inability to produce local estimates. Assumes that the grade and variability of panels are known and are correct. This also applies to any methodology utilised in which the known variability is unrepresentative of the actual variability.

DEFINITION OF MINERAL RESERVES

According to CRIRSCO, a mineral reserve is "the economically mineable part of a Measured or Indicated mineral resource demonstrated by at least a Preliminary Feasibility Study."

The study should include an assessment of the technical and economic factors that will influence the extraction and processing of the mineral, as well as the market conditions that will affect its sale. The reserve estimate should also be based on reasonable assumptions about the mine design, production rates, and costs.

CRITERIA FOR CONVERTING RESOURCES TO RESERVES

To convert mineral resources to reserves, the following criteria must be met:

- *Geological knowledge:* There must be a reasonable level of geological knowledge about the deposit, including its size, shape, and continuity. The deposit should be defined by a reasonable number of drill holes and samples.
- *Technical and economic viability:* There must be a preliminary feasibility study that demonstrates the technical and economic viability of the deposit. This study should include an assessment of the mining, processing, and infrastructure requirements, as well as the expected costs and revenues.
- *Mine design and production schedule:* There must be a preliminary mine design and production schedule that is based on the orebody model and considers the mining and processing methods, production rates, and costs.
- *Environmental and social considerations:* The environmental and social impacts of the mining project must be evaluated and managed, including the impacts on water, air, land, and biodiversity, as well as the impacts on local communities and stakeholders.
- *Market conditions:* The market conditions for the mineral product must be evaluated, including the demand, price, and supply of the product, as well as the transportation and marketing requirements.
- *Legal and regulatory considerations:* The legal and regulatory requirements for the mining project must be evaluated, including the permits, licenses, and approvals needed to operate the mine.
- *Risk assessment:* The risks associated with the mining project must be identified and evaluated, including the geological, technical, economic, environmental, and social risks.
- *Financial assessment:* The economic viability of the mining project must be evaluated, including the expected costs, revenues, profits, and return on investment.

CATEGORIES OF MINERAL RESERVES

CRIRSCO defines three categories of mineral reserves, based on the level of confidence in the estimate:

- *Proven Reserves:* These are reserves that can be mined with a high degree of confidence, based on the level of geological knowledge and the results of drilling and sampling. The reserve estimate should be based on a minimum of two or more drill holes that have been assayed.
- *Probable Reserves:* These are reserves that can be mined with a reasonable degree of confidence, based on the level of geological knowledge and the results of drilling and sampling. The reserve estimate should be based on a minimum of three or more drill holes that have been assayed.
- *Feasibility Study Reserves:* These are reserves that have been evaluated in a preliminary feasibility study and are economically mineable with a high degree of confidence. The reserve estimate should be based on a comprehensive analysis of the technical and economic factors that will influence the extraction and processing of the mineral, as well as the market conditions that will affect its sale.

REPORTING MINERAL RESERVES

CRIRSCO provides guidelines for reporting mineral reserves, which are intended to promote transparency and consistency in reporting. The guidelines include the following requirements:

- *Clear and concise reporting:* The mineral reserves should be reported in a clear and concise manner, using appropriate technical terms and units of measurement.
- *Materiality:* The mineral reserves should be reported only if they are material to the mining project, and if they have a reasonable chance of being extracted and sold.
- *Competent person:* The mineral reserves should be reported by a competent person who has the necessary qualifications and experience to assess the mineral resources and reserves.
- *Disclosure of assumptions and methods:* The assumptions and methods used to estimate the mineral reserves should be disclosed, including the parameters used in the estimation, such as the cut-off grade, recovery rate, and mining costs.
- *Sensitivity analysis:* The sensitivity of the mineral reserve estimate to changes in the assumptions and methods should be disclosed, including the impact of changes in the cut-off grade, recovery rate, and mining costs.
- *Updating of reserves:* The mineral reserves should be updated periodically, based on new information or changes in the parameters used in the estimation.
- *Review and verification:* The mineral reserves should be reviewed and verified by an independent expert or auditor, to ensure that they are accurate and reliable.

- *Public disclosure:* The mineral reserves should be publicly disclosed, in accordance with the legal and regulatory requirements of the country where the mining project is located.

ADDITIONAL CONSIDERATIONS

Here one is faced with two methods, the first being the one with the highest level of confidence, and the second method being that of obtaining the reserve by calculation only utilising the attributes assigned to the individual blocks of ground.

In the first case one does an actual mine plan, planning each piece of ground and scheduling it according to the period in which it becomes available to mine. Dip and strike stability pillars as well as bracket pillars are excluded from the schedule according to latest geological mapping and interpretation of structure as well as adhering to rock engineering principles and guidelines. This then becomes known as the Scheduled Reserve. Once the tonnages and grades have been obtained of the areas scheduled the additional reserve parameters are then applied these follow the ore flow as explained in the section under reconciliation. Here, the historical (not the planned) parameters such as mine call factor, tonnage discrepancy and plant recovery percentage are applied in the sequence encountered in the ore flow, producing the scheduled reserve from the estimated resource. Additionally, the reserve confidence (except in those exceptional cases mentioned previously) is obtained from the resource confidence, with a measured confidence translating into a proved reserve and an indicated confidence translating into a probable reserve. Inferred resources cannot and should not be reported as reserves. Once the areas under question have sufficient confidence due to additional sampling, etc., the reserve can be reconciled as being, for example, Probable reserves resulting from resource upgrade from Inferred to Indicated.

SPECIAL ADDITIONAL NOTES FOR CHAPTER 8: PLURIGAUSSIAN SIMULATIONS IN GEOSCIENCES

PluriGaussian simulations are powerful tools for geoscientists to model the spatial variability of geologic features, such as lithology, mineralisation, and fluid properties. This chapter introduces the mathematical foundations of PluriGaussian simulations and how they can be applied in geosciences for facies analysis and resource estimation uncertainty quantification and management.

BACKGROUND

In geosciences, the spatial variability of geological features is a critical aspect to understand, as it directly impacts the exploration, mining, and production of natural resources. PluriGaussian simulations are a class of geostatistical techniques that can model the spatial variability of geological features by explicitly considering the different patterns of spatial correlation within the data. This approach can be particularly useful in situations where the geological features are non-stationary or exhibit complex patterns of spatial correlation. PluriGaussian simulations are based on the theory of multiple-point statistics (MPS), which provides a framework for describing the spatial structure of geological features as a collection of patterns of different shapes and sizes. The MPS theory can be used to quantify the probability distribution of patterns within a training image, which is a two- or three-dimensional representation of the geological feature of interest.

MATHEMATICAL FOUNDATIONS

PluriGaussian simulations are based on a set of mathematical equations that describe the relationship between the observed data and the model of the spatial variability of the geological feature of interest. The equations are typically formulated in terms of the probability distribution of patterns within the training image and the correlation structure between different patterns.

The most used equation for PluriGaussian simulations is the truncated PluriGaussian model, which assumes that the spatial distribution of the geological feature of interest can be described as a mixture of different Gaussian distributions, each with a different mean and covariance matrix. The probability density function of the truncated PluriGaussian model can be written as:

$$f(x) = \Sigma j \; wj \, f\left(x \mid \mu j, \Sigma j\right) Ij(x)$$

where x is the spatial location of interest, wj is the weight of the jth Gaussian distribution, $f(x|\mu j, \Sigma j)$ is the probability density function of the jth Gaussian distribution with mean μj and covariance matrix Σj, and $Ij(x)$ is an indicator function that takes the value of one if x belongs to the jth pattern and zero otherwise.

The weights wj in the truncated PluriGaussian model can be estimated using maximum likelihood or Bayesian methods, based on the observed data and the assumed correlation structure between the different patterns. The correlation structure can be modelled using a variety of approaches, such as the covariance-based approach, which assumes that the correlation between patterns depends on their spatial distance, or the Markov random field approach, which models the correlation between patterns using a graphical model.

PROCESS OF PERFORMING PLURIGAUSSIAN SIMULATIONS

1. Define the geological problem: The first step is to clearly define the geological problem and identify the key geological features that need to be modelled.
2. Collect and analyse the data: The next step is to collect and analyse the available geological data, such as drill-hole data, outcrop data, and geophysical data. The data should be analysed to identify the statistical characteristics of the geological features of interest.
3. Define the geological constraints: Based on the analysis of the data, the geological constraints should be defined, such as variograms, spatial correlation models, and training images. The training image should be selected based on its ability to represent the geological features of interest.

4. Choosing the Model Type: The first step is to choose the appropriate PluriGaussian model type for the geological features of interest. This includes selecting the number of facies and defining the facies proportions.

5. Estimating the Parameter Values: The next step is to estimate the parameter values for each facies using geological data such as core samples, borehole logs, outcrop measurements, and geophysical surveys. These parameter values include the variogram range, sill, and nugget effect for each facies.

6. Generating Gaussian Values at Wells/Drill-Holes: Gaussian values are generated at wells or drill-holes using the estimated parameters from Step 5. These Gaussian values represent the local variability in the geological features.

7. Simulating Values at Grid Nodes Given Values at Wells: PSG simulations are performed to generate facies models at the desired resolution, such as grid nodes. The values at grid nodes are simulated using the Gaussian values generated at wells as input. The PSG algorithm includes different interpolation techniques, such as sequential Gaussian simulation (SGS), direct sequential simulation (DSS), and MPS.

8. Assessment the Simulation Results: The resulting facies models should be estimated using different statistical measures, such as cross-validation, entropy, and conditional simulations. The estimation should be used to quantify the uncertainty associated with the facies model and to identify areas where additional data may be required.

9. Using the Facies Model: The facies model can be used for a range of applications, such as resource estimation, mine planning, and risk management. The model should be used with caution, and its limitations and uncertainties should be clearly communicated.

10. Updating the Facies Model: The facies model should be updated as new data becomes available or as the understanding of the geological problem evolves. The PSG simulations can be repeated using the updated parameter values to generate new facies models.

BOX 1.1   CASE STUDY OF COPPER PORPHYRY MINERAL RESOURCE ESTIMATION IN SOUTH AMERICA USING PLURIGAUSSIAN SIMULATION

A mining company in South America was exploring a copper porphyry deposit, with the goal of estimating the mineral resources present. The deposit was characterised by complex geology, with several lithological units and structural features that affected the distribution of copper mineralisation. The company decided to use PluriGaussian simulation to model the spatial variability of the copper grades and estimate the mineral resources with a high level of confidence.

METHODOLOGY

Step 1: Choosing the Model Type

The geological model was divided into four facies: porphyry, breccia, clay alteration, and silicification. The proportions of the facies were estimated from the geological mapping and core samples.

Step 2: Estimating the Parameter Values

The variogram parameters were estimated for each facies using the core samples and geophysical surveys. The variogram models were chosen based on the geological knowledge and the quality of the data. The parameters were validated using a cross-validation approach.

Step 3: Generating Gaussian Values at Drill-Holes

Gaussian values were generated at drill holes using the estimated parameters from Step 2. The Gaussian values represent the local variability in the geological features.

Step 4: Simulating Values at Grid Nodes Given Values at Drill-Holes

The PSG simulations were performed using the Gaussian values generated at drill-holes as input. The simulations were run at a resolution of 20 m. The simulations were performed using the DSS method. The number of realisations was chosen to achieve a reasonable level of uncertainty reduction.

Step 5: Evaluating the Simulation Results

The simulations were evaluated using different statistical measures, such as cross-validation, entropy, and conditional simulations. The simulations were compared to the available geological information, such as drill-hole data, geological maps, and rock chip samples. The simulations were validated by comparing the simulated copper grades to the observed copper grades at drill-hole intersections.

Step 6: Using the Facies Model

The resulting facies models were used for copper mineral resource estimation. The estimation was performed using a block-model approach with 20 × 20 × 10 m blocks. The grade-tonnage curves were generated for each facies and the combined mineralisation. The mineral resources were estimated using a 0.2% copper cut-off grade.

Step 7: Updating the Facies Model

The facies model was updated as new data became available. The updated model was used to estimate the copper mineral resources using the same methodology as before.

RESULTS

The PluriGaussian simulation approach allowed for a more realistic representation of the geological complexity of the deposit. The resulting facies model was able to capture the spatial variability of the copper grades and the lithological units. The copper mineral resources were estimated to be 150 million tonnes at an average grade of 0.4% copper using a 0.2% copper cut-off grade.

CASE STUDY CONCLUSION

PluriGaussian simulation is a powerful tool for modelling the spatial variability of mineral deposits in complex geologic settings. By incorporating multiple geologic features and honouring the spatial correlation of the data, it allows for more accurate and robust mineral resource estimates. The case study presented here demonstrates how PluriGaussian simulation was used to estimate the copper porphyry mineral resources in a South American deposit, highlighting the importance of each step in the process.

CONCLUSION

In this chapter, we have explored into the realm of nonlinear geostatistical methods, which surpass traditional linear estimation approaches, to estimate recoverable mineral resources accurately. By employing these advanced techniques, we can model grade-tonnage relationships corresponding to mining selectivity, particularly in relation to the volume or support. Recoverable Reserves and resources are key components of any mineral project, with different types of estimates required at various stages of the mine life. Global reserves are necessary during the project development phase, while local reserves become important for detailed mine planning and pit/underground mine optimisation. The estimation of resources remains consistent before calculating reserves, but the latter necessitates economic evaluation and careful pit or underground planning. However, this estimation process becomes more intricate when considering equipment-related factors and the corresponding scale of production. Indeed, Recoverable Reserves are greatly influenced by the selectivity achievable with the equipment used. Larger mining equipment tends to be more productive, but this advantage comes at the cost of selectivity. Large equipment cannot differentiate between ore and waste as easily or accurately as smaller equipment. This selectivity issue is closely tied to the concept of the selective mining unit, which represents the smallest volume of material where ore-waste classification is determined. Therefore, the SMU becomes a clear measure of equipment selectivity and significantly impacts the reserve calculation. Given the practical significance of these methods for mine geologists involved in estimating recoverable resources and converting them into mineral reserves, we have provided sufficient detail on change-of-support techniques for their practical application. Specifically, the application of conditional simulations, including SGS, and special methods such as uniform conditioning and localised uniform conditioning, have emerged as a powerful tool for quantifying uncertainty and calculating recoverable mineral resources. SGS allows us to generate multiple realisations akin to Monte Carlo samples, representing potential outcomes within the space of uncertainty. These realisations enable a comprehensive picture of the model's uncertainty domain, facilitating a robust estimation of mineral reserves and guiding risk mitigation strategies. The significance of conditional simulations in ore deposit modelling cannot be overstated, as they enable mining professionals to move beyond single-point estimates and embrace the inherent variability and uncertainty present in geological data. By generating multiple realisations, mining companies can assess various scenarios and optimise resource management decisions, leading to improved operational efficiency and financial outcomes. Uniform Conditioning estimates the distribution of SMUs within each estimated panel. This means that given the grade of a large block (the estimated panel), we can calculate the distribution of SMUs within that panel. Recoverable Reserves are then calculated using the derived SMU distribution from the panel estimate and the change-of-support model. The discrete Gaussian model is employed to accomplish this change of support. However, the uniform conditioning method has some limitations. It does not provide any spatial location for the SMUs contained in each panel, and high-grade and low-grade SMUs could be located anywhere within the panel. Furthermore, two panels with the same estimated grade will yield the same reserves, regardless of surrounding data. To address these limitations, Localised Uniform Conditioning (LUC) comes into play. LUC predicts the spatial locations of economically extractable mineralisation by assigning a single grade to each SMU-sized block. By enhancing the conventional uniform conditioning approach through localisation, LUC refines the model results, providing more accurate spatial information. The grades of the SMUs are derived from the conventional uniform conditioning grade-tonnage relationships.

In the broader context, mining companies consider various factors such as mining and processing costs, market demand, and commodity prices to determine the economically viable portion of mineral resources that can be extracted profitably. Quantifying uncertainty across different categories of mineral reserves is also a crucial part of the estimation process, offering stakeholders a clear understanding of the reliability and variability of the estimates. Moreover, adhering to industry reporting standards, such as the Australian Joint Ore Reserves Committee (JORC) Code and Canadian Institute of Mining, Metallurgy and Petroleum (CIM) Definition Standards, ensures transparency and consistency in disclosing critical information on mineral reserves. This includes essential details such as tonnage, grade, classification, and methodologies used in the estimation process. Ultimately, quantifying uncertainty in reporting provides stakeholders with a realistic and comprehensive view of the potential risks and variability associated with the mineral reserves. Overall, nonlinear geostatistical methods have brought significant advancements to the estimation of mineral reserves. These techniques enable mining professionals to better account for uncertainty, variability, and the impact of equipment selectivity on reserve calculations.

REFERENCES

Abzalov, M. Z. (2006). Localised uniform conditioning (LUC): A new approach for direct modelling of small blocks. *Mathematical Geology*, *38*, 393–411.

Assibey-Bonsu, W. (1998). Use of uniform conditioning technique for recoverable resource/reserve estimation in a gold deposit. Proceeding of Geocongress '98, Geological Society of South Africa, pp. 68–74.

Chiles, J.-P. & Delfiner, P. (1999). *Geostatistics: Modelling spatial uncertainty*. New York: Wiley.

Deutsch, C. V. & Journel, A. G. (1992). *GSLIB: Geostatistical software library and user's guide*. New York: Oxford University Press.

Journel, A. G. & Kyriakidis, P. C. (2004). *Evaluation of mineral reserves: A simulation approach.* New York: Oxford University Press.

Krige, D. G., Assibey-Bonsu, W., & Tolmay, L. C. K. (2004). *Post-processing of SK estimators and simulations for assessment of recoverable resources and reserves for mining projects geostats.* Banf.

Matheron, G. (1974). Les fonctions de transfert des petits panneaux. Technical Report N-395, Centre de Géostatistique, France.

Neufeld, C. & Deutsch, C. V. (2005). Calculating Recoverable Reserves with uniform conditioning. *Proceedings of IAMG '05: GIS and Spatial Analysis, 2,* 1065–1070.

Remacre, A. Z. (1987). Conditioning by the panel grade for recovery estimation of non-homogeneous orebodies. In Matheron G. & Armstrong M. (eds.), *Geostatistical case studies. Quantitative geology and geostatistics* (Vol. 2). Dordrecht: Springer. https://doi.org/10.1007/978-94-009-3383-5_8

Rivoirard J. (1994). *Introduction to disjunctive kriging and non-linear geostatistics.* Oxford: Clarendon Press.

Wackernagel, H. (2003). Geostatistical models and kriging. *IFAC Proceedings Volumes, 36*(16), 543–548.

EXERCISE FOR CHAPTER 8

1. What are the CRIRSCO guidelines for reporting mineral reserves?

   Answer: The CRIRSCO guidelines include requirements for clear and concise reporting, materiality, competent person, disclosure of assumptions and methods, sensitivity analysis, updating of reserves, review and verification, and public disclosure.

2. Who should report mineral reserves according to the CRIRSCO guidelines?

   Answer: Mineral reserves should be reported by a competent person who has the necessary qualifications and experience to assess the mineral resources and reserves.

3. What is the purpose of the CRIRSCO guidelines for reporting mineral reserves?

   Answer: The purpose of the CRIRSCO guidelines is to promote transparency and consistency in reporting mineral reserves, to facilitate informed investment decisions and to ensure responsible mining practices.

4. What is materiality in the context of reporting mineral reserves?

   Answer: Materiality refers to the requirement that mineral reserves should be reported only if they are material to the mining project, and if they have a reasonable chance of being extracted and sold.

5. Why is it important to disclose the assumptions and methods used to estimate mineral reserves?

   Answer: It is important to disclose the assumptions and methods used to estimate mineral reserves to provide transparency and to allow for comparison and verification by investors, regulators, and other stakeholders.

6. What is sensitivity analysis in the context of reporting mineral reserves?

   Answer: Sensitivity analysis refers to the requirement that the impact of changes in the assumptions and methods used to estimate mineral reserves should be disclosed, including the impact of changes in the cut-off grade, recovery rate, and mining costs.

7. Why is the updating of mineral reserves important?

Answer: The updating of mineral reserves is important to ensure that the estimates are based on the latest available information, and to reflect changes in the parameters used in the estimation.

8. What is the role of independent experts or auditors in the reporting of mineral reserves?

Answer: Independent experts or auditors are responsible for reviewing and verifying the mineral reserves to ensure their accuracy and reliability.

9. What is the benefit of public disclosure of mineral reserves?

Answer: Public disclosure of mineral reserves provides transparency and allows for comparison and verification by investors, regulators, and other stakeholders, which promotes informed decision-making and responsible mining practices.

10. What are some of the factors that need to be considered when converting mineral resources to mineral reserves?

Answer: Factors that need to be considered when converting mineral resources to mineral reserves include the geology, technical and economic viability, market conditions, environmental and social impacts, legal and regulatory requirements, and risks associated with the mining project.

EXERCISE FOR PLURIGAUSSIAN SIMULATION

1. What is PluriGaussian simulation?

PluriGaussian simulation is a geostatistical method that enables the integration of multiple sources of geological information, such as core data, well logs, and geological maps, to simulate geological models. This method allows for a more realistic representation of the spatial variability of geological parameters.

2. Is Multi-Point Geostatistics (MPS) and PluriGaussian simulation the same?

No, Multi-Point Geostatistics (MPS) and PluriGaussian simulation (PGS) are not the same, although they are related.

MPS is a geostatistical technique that involves the simulation of a spatial pattern by conditioning on multiple data points instead of only a single point. MPS is based on the idea that geological heterogeneity is better represented by a set of patterns rather than a single variogram model. It involves using a training image, which is a 2D or 3D representation of the geological structure of interest, to generate multiple equiprobable realisations of the subsurface using a simulation algorithm. The training image can be a geological map, a seismic section, or any other type of image that captures the geological variability of the subsurface.

PluriGaussian simulation (PGS) is a specific type of MPS that uses multiple Gaussian distributions to represent the spatial variability of different geological facies. PGS involves dividing the subsurface into multiple facies and modelling each facies as a set of Gaussian distributions with different mean and variance values. The simulation algorithm then combines these

Gaussian distributions to generate equiprobable realisations of the subsurface that honour the spatial correlation between the different facies.

3. What is facies analysis?

Facies analysis is the process of identifying and characterising distinct geological units within a rock formation based on their lithological, petrophysical, and geometrical properties. These units are commonly referred to as facies.

4. How can PluriGaussian simulations be used in facies analysis?

PluriGaussian simulations can be used to simulate the spatial distribution of facies within a rock formation by integrating different sources of geological information. This allows for the creation of realistic geological models that can be used for a variety of purposes, such as mineral exploration, geotechnical analysis, and hydrogeological modelling.

5. What is resource estimation uncertainty quantification?

Resource estimation uncertainty quantification is the process of assessing the uncertainty associated with the estimation of mineral resources. This is important for making informed decisions about the economic viability of a mining project.

6. How can PluriGaussian simulations be used for resource estimation uncertainty quantification?

PluriGaussian simulations can be used to generate multiple realisations of geological models, which can be used to quantify the uncertainty associated with the estimation of mineral resources. By simulating a range of geological scenarios, it is possible to identify the most likely resource estimates and the associated uncertainty.

7. What is resource estimation management?

Resource estimation management is the process of managing the uncertainty associated with the estimation of mineral resources throughout the life of a mining project.

8. How can PluriGaussian simulations be used for resource estimation management?

PluriGaussian simulations can be used to update geological models based on new information, such as drilling results or additional geological data. This allows for more accurate resource estimates and better management of the uncertainty associated with these estimates.

9. What are some advantages of using PluriGaussian simulations?

PluriGaussian simulations allow for the integration of multiple sources of geological information, resulting in more realistic geological models. They also provide a way to quantify the uncertainty associated with resource estimates and can be used to update models as new information becomes available.

10. What are some challenges associated with using PluriGaussian simulations?

PluriGaussian simulations can be computationally intensive, particularly when simulating complex geological models. They also require careful consideration of the appropriate geological parameters to include and how they should be weighted.

11. What is the future of PluriGaussian simulations in geosciences?

PluriGaussian simulations are expected to continue to play a significant role in geostatistical modelling in the future, particularly as more sophisticated algorithms and computing power become available. They are likely to be used for a range of applications, including mineral exploration, geotechnical analysis, and hydrogeological modelling, and will help to improve our understanding of the spatial variability of geological parameters.

9 Grade control in mineral resources management and mine planning

INTRODUCTION

Grade control is an indispensable and intricate process at the heart of mining operations. The overarching goal is to enhance the mining project's overall value while mitigating potential risks. This key activity revolves around delivering materials of the utmost quality and optimal grade to the mineral processing plant for beneficiation, achieved through the accurate and precise delineation of ore and waste to ensure profitability. Central to the success of any grade control programme is the meticulous collection of high-quality samples within a well-defined geological context. These samples act as the foundation upon which the entire mining operation is built, providing essential insights and data that drive critical aspects of mine planning and decision-making.

One of the primary roles of sampling is to inform grade resource estimation, allowing mining professionals to understand the mineral deposits' distribution, quantity, and quality. This knowledge is invaluable for estimating the mineral reserve, planning mining activities, and optimising the extraction process. Accurate geological resource estimation empowers mining companies to make informed financial projections and develop sustainable, long-term strategies for their operations. Grade control relies heavily on the data acquired through sampling. By continuously monitoring the grade of extracted material, mining operations can adjust their processes in real-time to maintain a consistent and optimal feed to the mill. This ensures the extraction of the highest-grade ore, which maximises the operation's profitability.

Beyond financial considerations, sampling also plays a key role in addressing geo-environmental concerns. By assessing the composition and potential contaminants of the extracted material, mining companies can implement environmentally responsible practices and reduce their ecological footprint. This commitment to sustainable mining practices fosters positive relationships with local communities and regulatory bodies and safeguards the surrounding environment for future generations. Sampling facilitates geometallurgical assessments, a critical aspect of mining that analyses the interaction between the ore's geological characteristics and the metallurgical processes. This insight enables mining engineers and metallurgists to optimise the mineral processing techniques, enhancing recovery rates, and minimising energy consumption. As a result, mining operations become more efficient, economically viable, and environmentally sustainable.

In grade control, holes are drilled near each other to check the quality and grade of the material while mining is in progress. This is typically done to help define the short-term mine plan. In open-pit operations, blast-hole cutting samples are collected

 DOI: 10.1201/9781032650388-9

by down-the-hole hammer drills, and bench Reserves are classified as ore, low-grade material, waste material, or various metal types. Some grade control strategies also involve sampling truck and shovel loads to ensure that material is assigned to the correct stockpiles or waste dumps. The grade control process may involve mapping and sampling stope faces or benches, tram carloads and draw-point muck piles, broken rock at a recently blasted face, jackhammer cuttings or diamond drill cores (Malisa and Genc, 2020). A geologist's job is to ensure that production closely follows mineralised zones and to minimise dilution during the mining process. In this way, grade control ensures that the material being fed to the mill is of economic grade and that large fluctuations in grade are minimised by blending ore from different benches or different sources (Ruiseco et al., 2016). The feed to the mill and concentrator must be maintained as close to the original plan design specifications as possible.

DRIVERS FOR EFFECTIVE GRADE CONTROL

There are several factors that can drive the need for effective grade control in mining operations. Here are ten drivers for grade control and an explanation of each:

- Economic viability: One of the most important drivers for grade control is economic viability. The value of the minerals being mined must be higher than the costs of extraction and processing for the mine to be profitable. By controlling the material grade that is processed in the plant to above the economic cut-off, the mine can ensure that, all other factors being equal, it will make a profit.
- Market demands: Another driver for grade control is the demands of the market. Different industries and applications may require different grades, or qualities, of minerals, with penalties being levied on the mine if the delivered material is not within the required, contractually defined, specifications.
- Environmental regulations: Environmental regulations can also drive the need for effective grade control. Mines must comply with regulations that limit the amount of waste and pollution that can be generated during the mining process. By controlling the grade of the ore, the mine can minimise waste and reduce the environmental impact of the operation.
- Safety: Safety is another important driver for grade control. The mining process can be dangerous and ensuring that the ore is of a consistent quality can help to minimise the risks associated with mining (by only mining and processing economic materials).
- Production efficiency: Controlling the grade of the ore can also help improve production efficiency. By ensuring that the material being mined is of a consistent quality, the mine can optimise its processing operations, hereby reducing the amount of time and energy required to produce the end-product.
- Resource management: Effective grade control can also help to manage resources more effectively. By accurately determining the grade of the ore, the mine can allocate resources more efficiently, reducing waste and maximising the value of the mineral deposit.

- Equipment utilisation: By controlling the grade of the ore, the mine can also optimise the utilisation of its equipment. Equipment is expensive to operate and maintain, and effective grade control can help ensure that the equipment is used as efficiently as possible.
- Mineral recovery: The recovery of minerals is another important driver for grade control. By ensuring that the ore is of a consistent quality, the mine can optimise its recovery processes, reducing losses and maximising the amount of material that can be sold.
- Mine planning: Effective grade control can also help with mine planning. By accurately determining the grade of the ore, the mine can plan its operations more effectively, reducing the risk of unexpected issues arising during the mining process.
- Investor confidence: Finally, effective grade control can help to build investor confidence. Investors want to see that a mine is being managed effectively and that the material being produced is of consistent, high quality. A mine can build trust and attract new investors by demonstrating good grade control practices.

WORKFLOW FOR GRADE CONTROL

The main steps and components that are typically included in a flowsheet for good grade control in mining operations include (Figure 9.1):

- Sampling: The first step in good grade control is correct sampling. The mine must ensure that it takes representative samples of the ore from the mine face or the process stream. This requires careful planning and execution to minimise errors and biases in the sampling process.
- Sample Preparation: Once the samples are collected, they must be prepared for analysis. This may include crushing, grinding (pulverising), and homogenising the sample to ensure it is a representative subset of the mined material.
- Analysis: The prepared sample is then analysed to determine its grade. This may involve various analytical techniques, such as fire assay, X-ray fluorescence (XRF), atomic absorption spectroscopy (AAS), or inductively coupled plasma (ICP) analysis.
- QA/QC: Measures must be implemented to ensure that the analysis is accurate and precise. This will include the insertion of blanks, duplicates, and standards along with the material that has been collected for assaying purposes, to verify the accuracy and precision of the analysis.
- Data Management: The results of the analysis must be carefully recorded and managed to ensure that they are used effectively. This may involve using software systems to track sample locations, analysis results, and other relevant data.
- Grade Control Block Model: Once the analysis results are available, the mine must create a grade control block model using Kriging or other suitable interpolation techniques. The block model should be created at the appropriate block size, such as the selective mining unit (SMU) size, to capture the variability of the ore. The block model can be used to determine the grade of the ore in different parts of the mine and to guide mining and processing operations. This may involve adjusting mining parameters, such as blast hole size and spacing,

to ensure that the ore is mined at the desired grade. The block model can also be used to optimise processing operations, such as adjusting the crushing and grinding circuits to match the grade of the ore being processed. Effective grade control block modelling requires careful planning and execution to ensure that the model accurately reflects the variability of the ore and that it is being used effectively to guide mining and processing operations.

- Interpretation: The final step in good grade control is interpretation. The mine must interpret the results of the analysis and use this information to make informed decisions about mining and processing operations. This may involve comparing the grade of the ore to the mine plan, adjusting mining and processing parameters, and communicating the results to relevant stakeholders.

By following these steps and ensuring that each component of the flowsheet is executed effectively, the mine can achieve good grade control and optimise its mining and processing operations.

THE SIGNIFICANCE OF SAMPLING IN GRADE CONTROL

Sampling plays an indispensable role in every aspect of the mining process, making it a critical component throughout the entire mine value chain. Given the complexity of mining environments, attempting to analyse all the material in advance would be an overwhelming and impractical task. Instead, sampling, which involves carefully selecting representative portions of in-situ and broken materials, serves as the foundation for acquiring invaluable information used in various stages of mining operations. One of the key areas where sampling holds immense importance is in geological and grade control models. Accurate resource estimation is essential for determining the economic viability of a mining project and planning the most efficient extraction methods. Without reliable data from sampling, making informed decisions would become nearly impossible. The implications of sampling errors can be profound, with repercussions both on financial outcomes and intangible losses. When samples fail to represent the mineralisation in a deposit correctly, it can result in misleading resource estimates and misguided mining decisions. Inaccurate information about the ore body's grade and distribution can result in costly mistakes, inefficient extraction practices, and even premature closure of mining operations.

This issue becomes particularly pronounced in deposits where coarse gold particles dominate, as standard sampling methods may not capture the true distribution of these valuable minerals. Consequently, specialised protocols and sampling techniques may be necessary to ensure accurate results in such scenarios. Failure to obtain representative samples can lead to the undervaluation of the deposit's mean grade. This undervaluation, in turn, can lead to the overestimation of block grades below the economic cut-off value, which could erroneously classify valuable ore blocks as waste. As a result, economically viable resources might remain untapped, leading to missed opportunities and reduced profitability for the mining company. Implementing a robust QA/QC programme is crucial to safeguard data integrity and ensure the accuracy of sampling results. This programme involves the establishment of well-documented procedures, stringent sample security measures, and continuous monitoring of precision, accuracy, and contamination risks.

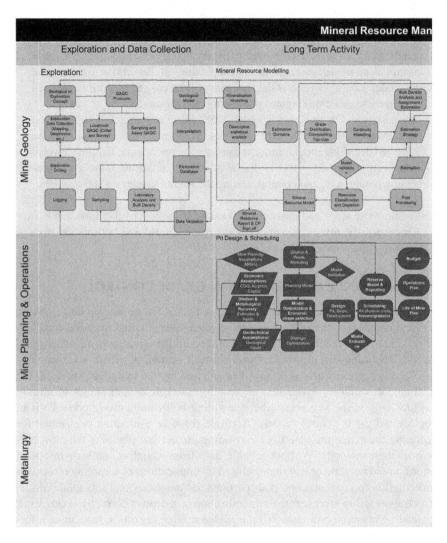

FIGURE 9.1 Illustration of grade control (GC) process workflow. The main grade control activities are in light boxes. (For a larger version of this figure see the Support Material at www.routledge.com/9781032599267.)

In-depth documentation of all sampling processes and protocols is essential for traceability, enabling thorough audits and verifications. Stringent sample security measures ensure that samples are not tampered with or contaminated, guaranteeing the reliability of the data collected. Continuously monitoring precision and accuracy helps identify and rectify any potential issues before they can escalate and impact decision-making. The QA/QC programme also encompasses regular checks for contamination, ensuring that external elements do not influence the sample's composition. This is particularly important when dealing with high-value commodities such as gold, where even minute contamination can significantly impact the accuracy of results. By adhering steadfastly to these QA/QC practices, mining operations can instil confidence in

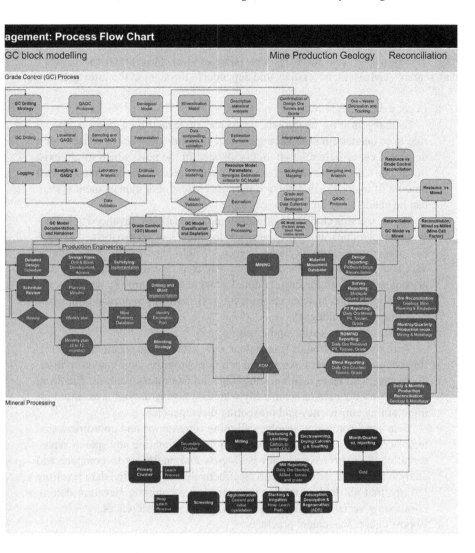

the reliability and accuracy of the data they collect. Trustworthy and precise data are fundamental to successful mining ventures, enabling companies to optimise their operations, maximise profitability, and minimise potential risks and uncertainties.

DATA MANAGEMENT IN GRADE CONTROL: ENSURING DATA INTEGRITY

Maintaining data integrity is a critical aspect of successful grade control in mining operations. Accurate and reliable data is the foundation for making informed decisions that affect the profitability and efficiency of the entire mining process. This section explores the significance of rigorous data management practices in grade control and emphasises the importance of incorporating both legacy data and new data synchronisation and levelling. The integration of these practices ensures that data remains consistent, accessible, and dependable throughout the mining lifecycle.

a. Legacy data integration:

In many mining operations, historical or legacy data plays a crucial role in understanding the geological and mineralogical characteristics of a deposit. Legacy data refers to data collected in the past, which might exist in various formats, databases, or even on paper records. Integrating this vast amount of historical information into the current data management system can be challenging but immensely valuable.

To address this challenge, mining companies employ advanced data integration techniques and tools. Data cleaning processes are performed to identify and correct errors or inconsistencies within legacy data. Additionally, data standardisation methods are implemented to ensure that legacy data are compatible with the current data format and structure. By doing so, the mining operation can leverage valuable insights from historical data while maintaining data integrity.

b. New data synchronisation and levelling:

With the integration of legacy data, new data synchronisation and levelling become paramount. As mining activities progress, new data is continuously generated from drilling, sampling, and laboratory analysis. Ensuring that this new data aligns seamlessly with the existing dataset is essential for maintaining consistency and preventing discrepancies.

Data synchronisation involves aligning timestamps and coordinate systems across various datasets to create a unified timeline and spatial representation. This process enables geologists and engineers to compare and analyse data from different sources effectively. Additionally, data levelling ensures that all data points are referenced to a common elevation datum, eliminating variations caused by differing elevation references.

c. Robust data management practices:

To achieve data integrity, mining operations must implement robust data management practices. Some key components include:

i.   Documentation: Comprehensive and well-organised documentation is essential. It includes detailed information about sampling locations, methodology, equipment used, and personnel involved. Proper documentation enables easy tracing of data sources and supports audit trails.

ii. Sample security: Implementing strict protocols for sample handling and storage is crucial. Proper labelling, tracking, and secure storage of samples prevent contamination, loss, or mix-up, which could lead to inaccurate data.

iii. Quality Assurance/Quality Control (QA/QC) Measures: Rigorous QA/QC procedures are fundamental in identifying and rectifying errors. Regular checks for precision and accuracy ensure that the data collected is consistent and reliable.

iv. Data Auditing: Regular audits of the data management system help identify potential weaknesses or areas for improvement. External auditors can also be engaged to provide an unbiased assessment of data integrity.

d. Real-time data monitoring:

With the advent of advanced technologies, real-time data monitoring has become feasible in mining operations. Implementing automated data capture and monitoring systems ensures that data is continuously updated and validated. This real-time monitoring enhances the ability to detect and address data quality issues promptly.

LOCALISED RESOURCE ESTIMATION FOR GRADE CONTROL MODEL

Localised resource estimation for grade control is a important process in mining operations, as it directly influences the decision-making related to ore and waste material extraction. Accurate grade control models are essential for optimising mining processes, minimising waste, and maximising the recovery of valuable ore. This section explores the significance of high-quality data inputs, the complexities of model interpretation, and the crucial role of competent persons in ensuring reliable and informative grade control models.

a. Importance of high-quality data inputs

Developing accurate grade control models heavily relies on the quality of data inputs. These data inputs are derived from well-designed and executed sampling programmes. The sampling process should consider factors such as spatial distribution, sample size, and representativeness of the samples. To achieve reliable localised resource estimation, the following aspects are crucial:

i.   Spatial distribution: Sampling points should be distributed adequately throughout the stope or bench to capture the variability of the orebody. A robust spatial sampling design ensures a representative sample set, reducing bias in the estimation process.

ii. Sample size: The sample size must be appropriate to capture the geological heterogeneity within the target area. Insufficient sample size can lead to unreliable estimates, while an excessively large sample size may result in unnecessary costs.

iii. Sample representativeness: Samples must accurately represent the orebody's characteristics. Techniques such as core logging, geostatistical analysis, and geophysical surveys can help identify and address sample representativeness issues.

b. Complexities of model interpretation

Interpreting grade control models is a complex task that demands a thorough understanding of various factors. Geologists and engineers responsible for model interpretation must consider the geological context, data quality, and uncertainties associated with the estimation process. Several challenges arise during model interpretation:

i.   Geology and structure: Understanding the geological features and structural complexities is critical for interpreting grade distribution accurately. Faults, folds, and lithological variations can significantly influence the spatial distribution of mineralisation.

ii. Data quality: The accuracy and reliability of the grade control model depend on the quality of the data inputs. Geologists must identify potential errors and uncertainties and apply appropriate adjustments during the interpretation process.

iii. Sampling errors: Inherent errors associated with the sampling process can impact the reliability of the grade control model. Understanding the magnitude and nature of these errors is crucial for making informed decisions based on the estimation results.

c. Competent persons and decision-making

The responsibility of interpreting grade control models falls on competent persons with expertise in geology, resource estimation, and mining practices. Competent persons must critically assess the model's output, considering the geological context and data quality. Their expertise ensures that the interpretation aligns with the objectives of the mining operation and that the results can be confidently used for decision-making.

The decisions made based on the grade control model interpretation directly affect mining efficiency, material extraction, and waste management. It is imperative to have a clear understanding of the model's accuracy and reliability before implementing any operational changes.

PROCESSES FOR GRADE CONTROL

It is essential to ensure that the tonnage and grade estimated to be mined and sent to the mill correspond to the actual tonnage and grade received. Achieving such correspondence requires observation, monitoring, and timely actions. There may be a mismatch between what is needed and what is recovered from plants for the following reasons:

i. Higher grade ore being trammed as waste or transported to low-grade or waste stockpiles.

ii. Lower grade (subeconomic) or waste grade material being trammed as ore and processed.

iii. Excessive spillage when transporting ore.

iv. Excessive water use results in loss of fine material (generally high-grade material) during transportation of ore.

v. Excessive water use results in ore loss into the footwall (a specific problem in underground mines).

vi. Excessive fragmentation, resulting in increased liberation of minerals of interest, and hence resulting in increased risk of mineral loss due to leakage, leaching, or oxidisation.

vii. Ore being left behind in underground workings due to poor cleaning practices.

viii. Ore being left behind in the footwall, hanging wall, or not being fully exposed (a specific problem in underground operations).

ix. Unplanned sterilisation of ore, due to the presence of unanticipated geological features.

x. Incorrect identification of the correct horizon to mine (Yes, it happens).

xi. Excessive overbreak in underground mining operations, thus diluting the ore grade, by increasing the tonnages of rock delivered for processing.

xii. The area that was planned to be mined not the same as the area that was mined.

xiii. The area that was planned is not mined at all.

xiv. Mining occurs in areas that are not part of the mine plan.

xv. Mining more than what was planned to be extracted from an area. This could result in sterilisation of adjacent mining areas if high and low grades ore was to be blended, alternatively, if the excessive mining occurs in sub economic ore, this will result in higher processing costs (reagent for example) and lower or no profit being realised.

xvi. Mining less than planned in an area. This could have an adverse effect on the mine plan, if this was a high-grade area.

xvii. Using backfill and backfilling above broken ore (a problem in underground operations).

xviii. Using backfill and having excessive backfill spillage, thus reducing grade of ore with backfill.

xix. Using a particle size greater than the liberation size, either intentionally for backfill, or unintentionally by substandard milling practices.

xx. Excessive conveyor belt spillages that are not cleaned up in a timely manner. This is not an 'actual' loss, as the material will enter the plant at some stage; however, it may not be allocated to the correct period. Alternatively, the ore may become oxidised and harder to recover.

xxi. Conveyor belt spillages are washed into the area outside the plant and then ignored.

xxii. Incorrect weightometer readings may occur due to the instrument being irregularly serviced or having service intervals exceeding makers' specifications.

There are several items on the list that require continuous real-time action. It is also noteworthy that no estimation methods are mentioned here. This is because grade control is about controlling the attributes that determine the actual grade of ore that will be processed in the plant. Often in the authors' experience, rigorous grade control measures are ignored, as they are regarded as needless, additional costs.

Consequently, the actual loss of content is ignored and never recovered, resulting in a loss of revenue. Here, the authors are trying to make the point that hiring someone to keep track of, monitor, and report on the points pertaining to workings and plant is well worth the investment.

CONCLUSION

Sampling is the cornerstone of grade control in mining operations. Accurate and representative samples within a geological context form the foundation for making informed decisions throughout the entire mining value chain. By implementing the Theory of Sampling and stringent QA/QC practices, mining companies can minimise errors and ensure the reliability and integrity of their data. Local resource estimation models for grade control aid in optimising mining operations, allowing for efficient material extraction and waste management. The interpretation of these models requires expertise and a thorough understanding of potential uncertainties. Mining companies must recognise the importance of sampling and its impact on the success and profitability of their operations. By prioritising high-quality sampling practices, adherence to Theory of Sampling (TOS) principles, and robust QA/QC measures, mining operations can confidently make critical decisions and maximise their value while reducing risks.

REFERENCES

Malisa, M. T. & Genc, B. (2020). Mine planning and optimisation techniques applied in an iron ore mine. In Topal, E. (ed), *Proceedings of the 28th international symposium on mine planning and equipment selection-MPES*, 2019, 28 (pp. 103–110). Cham: Springer International Publishing. https://doi.org/10.1007/978-3-030-33954-8_11.

Ruiseco, J. R., Williams, J., & Kumral, M. (2016). Optimizing ore–waste dig-limits as part of operational mine planning through genetic algorithms. *Natural Resources Research*, *25*, 473–485.

Spangenberg, I. C. (2012). *The status of sampling practice in the gold mining industry in Africa: Working towards an international standard for gold mining sample practices*. University of the Witwatersrand, School of Mining Engineering.

GRADE CONTROL EXERCISE QUESTIONS AND ANSWERS

1. What is the purpose of grade control in mining operations?

   Answer: The purpose of grade control in mining operations is to ensure that the ore being mined is of the desired grade and quality and that it is being extracted and processed efficiently.

2. What are some of the key drivers for good grade control in mining operations?

   Answer: Some of the key drivers for good grade control in mining operations include correct sampling, sample preparation, analysis, quality control, data management, interpretation, and grade control block modelling.

3. Why is it important to use representative sampling techniques in grade control?

   Answer: It is important to use representative sampling techniques in grade control to ensure that the samples being analysed accurately reflect the composition of the ore being mined. This helps to minimise errors and biases in the sampling process.

4. What analytical techniques are commonly used in grade control?

   Answer: Common analytical techniques used in grade control include fire assay, X-ray fluorescence (XRF), atomic absorption spectroscopy (AAS), or inductively coupled plasma (ICP) analysis.

5. What is the purpose of quality control in grade control?

   Answer: The purpose of quality control in grade control is to ensure that the analytical results are accurate and precise. This will include the insertion of blanks, duplicates, and standards along with samples being submitted for assaying purposes, to verify the accuracy and precision of the analysis.

6. How can grade control block modelling be used to optimise mining operations?

   Answer: Grade control block modelling can be used to optimise mining operations by guiding the selection of mining parameters, such as blast hole size and spacing, to ensure that the ore is being mined at the desired grade.

7. What is the selective mining unit (SMU) size, and why is it important in grade control block modelling?

   Answer: The selective mining unit (SMU) size is the block size used in grade control block modelling. It is important in grade control block modelling because it captures the variability of the ore and allows for more precise control of mining and processing operations.

8. What are some of the challenges associated with grade control in open pit mining operations?

   Answer: Some of the challenges associated with grade control in open pit mining operations include the difficulty of accurately characterising the orebody, the high variability of the ore, and the need to optimise mining operations while minimising costs.

9. What are some of the benefits of effective grade control in mining operations?

   Answer: Some of the benefits of effective grade control in mining operations include increased efficiency and productivity, improved recovery rates, reduced waste and rehandling, and enhanced profitability.

10. What is the role of geostatistics in grade control?

    Answer: Geostatistics is a key tool used in grade control to analyse and model the spatial distribution of the ore. This helps to optimise mining and processing operations by providing a more accurate picture of the orebody and its variability.

11. A mining company is conducting grade control sampling using blast hole drilling at 10-m intervals. The drill holes have a diameter of 127 mm and a depth of 30 m. If the bulk density of the ore is 2.5 g/cm³, what is the estimated weight of the sample collected from each blast hole?

    Answer: The volume of the sample collected from each blast hole can be calculated using the formula for the volume of a cylinder: $\pi \times (\text{diameter}/2)^2 \times$ depth. Plugging in the given values, we get:

    $$\text{Volume} = \pi \times (127/2)^2 \times 30 = 1{,}130{,}973 \text{ mm}^3 \text{ or } 1.13 \text{ L}.$$

    The weight of the sample can be calculated by multiplying the volume by the bulk density: 1.13 L × 2.5 g/cm³ = 2.83 kg. Therefore, the estimated weight of the sample collected from each blast hole is 2.83 kg.

12. A mining company is using a handheld XRF analyser to analyse grade control samples. The analyser has an accuracy of ±2% and a precision of ±1%. If a sample is found to contain 5.0% copper, what is the range of values that can be expected from the analyser?

    Answer: The range of values that can be expected from the analyser can be calculated by applying the accuracy and precision specifications. The expected range is calculated as follows:

    Accuracy range: 5.0% ± 2% = 4.9%–5.1%
    Precision range: 5.0% ± 1% = 4.95%–5.05%

    Therefore, the expected range of values from the analyser is 4.9% to 5.1% based on accuracy, and 4.95%–5.05% based on precision.

10 Mine-to-mill reconciliation for efficiency, resource utilisation, and profitability

INTRODUCTION

Mine-to-mill reconciliation is a process used in the mining industry to compare the actual production data from the mine with the expected production data at the mill. The purpose of this process is to identify the factors that cause differences between the actual and expected production data, and to take corrective actions if necessary. In this chapter, we will discuss the importance of mine-to-mill reconciliation, the steps involved in the process, and some of the challenges associated with it. Regular reconciliations will be required between the estimated stope grades, the grades indicated from stope/truck sampling and those reported by the mill. It is essential that this is undertaken so that modifications can be made to sampling practice or to the methods or parameters used to calculate grade, tonnage or contained metal. The mine-to-mill reconciliation takes into consideration both grade control and estimation. It follows the mine ore flow, taking cognisance of each aspect individually, and reconciling the individual impact of these.

Mine-to-mill reconciliation is an important process for several reasons. First, it allows for the detection of differences between the actual and expected production data, which can help to identify areas for improvement in the mining operation. Second, it provides a means for verifying the accuracy of the mine and mill data, which is important for the financial reporting of the mining company. Further, mine-to-mill reconciliation can help to improve the understanding of the ore characteristics and variability, which can inform the design and optimisation of the mining and processing operations.

STEPS IN MINE-TO-MILL RECONCILIATION

The following are the general steps involved in mine-to-mill reconciliation:

- Step 1: Data collection is the first step which involves collection of relevant data, including the actual production data from the mine and the expected production data at the mill. These data can include tonnages, grades, metal recoveries, and other parameters.

DOI: 10.1201/9781032650388-10

- Step 2: Comparison of actual and expected data involves comparing the actual and expected production data to identify any differences. This can also involve the use of statistical methods, such as variance analysis or regression analysis, to identify the factors that contribute to the differences.
- Step 3: Identification of causes of differences is the process used to identify the causes of the differences between the actual and expected production data. This can involve a detailed analysis of the ore characteristics, mining and processing methods, equipment performance, and other factors that may affect production.
- Step 4: Implementation of corrective actions is when corrective actions are taken to address the identified causes of the differences. This may involve changes to the mining or processing methods, equipment upgrades or replacements, or other measures to improve the efficiency and effectiveness of the operation.
- Step 5: Monitoring and Review are very important as they allow monitoring and reviewing the effectiveness of the corrective actions, and to make further adjustments as necessary. This may involve ongoing data collection and analysis, as well as regular reviews of the mining and processing operations to identify areas for improvement.

THE PROCESS FOR MINE-TO-MILL RECONCILIATION

The ore flow is the primary tool in reconciling the flow of tonnage and grades from mining through to the final product. The ore flow sheet is nothing magical or complicated; all it encompasses is how the ore gets from where it is blasted to the plant. At each stage of the ore's journey, one asks what can be determined at that stage and then inserts the information available at that stage. When two different sources of information do not agree, then one can compare the extent of the disagreement to best practice or, alternatively, once best practice has produced a historical figure that can be relied upon to be best practice in reality. This figure then becomes the standard for comparison. Utilising this method, one obtains a deviation from the expected at a certain position of the ore flow. This limits the areas that need to be examined to reconcile the reasons for the deviation to those areas that impact the portion of the ore flow under consideration. The method employed for a mine-to-mill reconciliation is as follows:

i. Create an ore flow that emulates the ore flow on your mine (see the example in Figure 10.1 part I and II)
ii. At each section, ask the question. What is it that I can determine at this stage?
iii. Now look at each stage and the corresponding previous stage and ask the question as to what you can compare at that stage. This may be tonnage only, or alternatively, content only or both tonnage and content.
iv. Calculate the difference between the two stages on the available parameter and compare the deviation from 100% to the acceptable standard expected on a monthly, six-monthly, or annualised basis. Remember that one will find that there will be greater variability monthly than on longer-term

comparisons. In some mines, this factor is mitigated by considering a three-
or six-month moving average to determine whether the deviation is of an
acute or chronic nature (Figure 10.2).

Let us now consider the previous ore flow example. We begin with ore from stoping and ore from sources other than stoping. In response to point two's question, the
answer is that the volume is mined for the specific period under consideration, as
well as the tonnage if the relative density is known. Furthermore, because the survey
normally measures the shape of the volume measured, this shape can be used against
the mineral resource estimation model to determine the *in-situ* value of the volume
mined. Thus, one can then obtain the corresponding estimate of the mineral content.
A similar set of operations can be accomplished for ore mined in development.

As the other sources come to a halt, careful consideration must be given to whether
mining the footwall or hanging wall is waste, or an alternative source of the mineral
of interest. Here all attempts must be made to sample these additional sources and
include them in the call. Ignoring these can and will lead to the compensatory nature
of the uncalled-for content hiding unsavoury mining practices that should be brought
to light. The next process along the way is the tramming of these sources. Here a
hopper factor may be utilised to monitor ore's progress from the stoping to the ore
passes at the shaft.

Although this is by no means a comprehensive measure, it can be utilised to either
rule out or include inspections for tonnage excesses or shortfalls that are found excessive in nature. Here there is a finer detail as locos may be exclusively assigned to
individual stopes or sources. One cannot reconcile grade and content at this stage of
the process. The next stage is hoisting. Here one has a higher level of granularity as
one can only determine tonnage based on the entire shaft. However, having said this,
an additional check on where things may be going wrong can be obtained for the
shaft by utilising a skip factor and thus getting a rough estimate of tonnage hoisted.

> Due diligence must be taken in obtaining a representative estimate of percent-
> age moisture in order that accurate tonnage may be obtained.

Here one can observe if the ore is being trammed as waste or waste as ore as the
proportions will not agree with those trammed.

The next stage examines where the tonnage hoisted is transported to. Here very
often, a weightometer may be employed, which should provide greater tonnage split
accuracy than the skip factor if regularly serviced. Nevertheless, this can be utilised to
verify incorrect tramming, as shown by the reef and waste tonnage hoisted and increase
the accuracy of the skip factor for future measurements. This will decide on whether it
would be a worthwhile exercise to take a trip to the waste rock dump to search for signs
of ore being sent to waste. At this stage, one also determines after sorting how much
tonnage goes to which stockpile (assuming the mine even has a stockpile).

Additionally, one normally utilises belt sampling at this stage to obtain an estimate of grade and content going to each source; the quality of the estimate will be

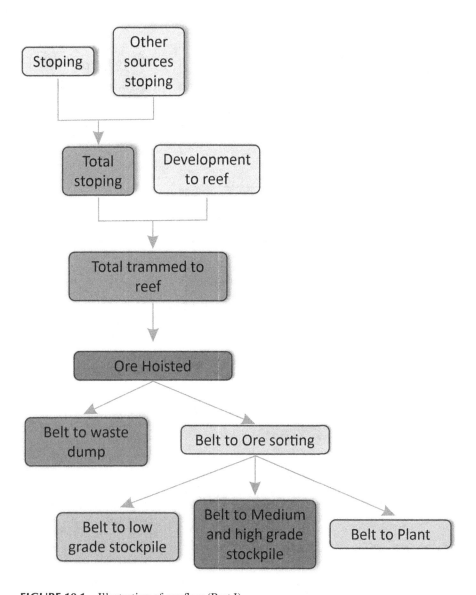

FIGURE 10.1   Illustration of ore flow (Part I).

determined by the method of belt sampling and whether the sampling occurs at a regular tonnage interval. The next stage is the sources sent to plant; here one may obtain tonnage from the low-grade, medium-grade, or high-grade stockpiles, as well as ore sent directly to the mill. How one assigns a grade to these sources depends upon:

a. Where and what method the tonnage is selected?
   i. A single stockpile in which ore is piled on top of the existing pile. Here one would utilise the average grade sent to the stockpile for that

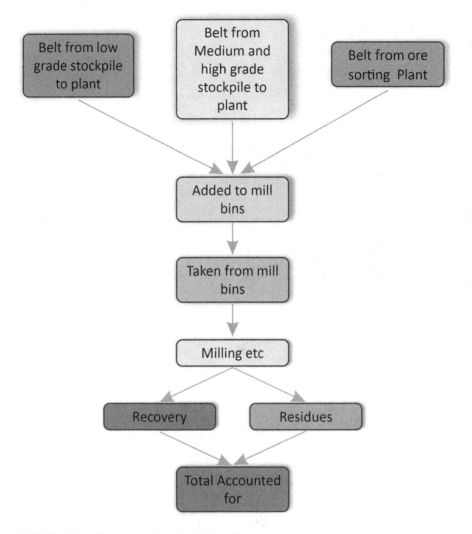

FIGURE 10.2   Illustration of ore flow (Part II).

month or the last month at which tonnage had been added. The tonnage and grade remaining on the stockpile are reconciled by adding the tonnage at the grade for that month and removing tonnage at a similar grade.

ii. Multiple stockpile stashes in which the position of the stockpile is noted in relation to the month deposited. Here one has a greater control of grade to mill, as if the grade is lagging, one can utilise a higher-grade portion, and alternatively, if the grade to mill is high, advantage of this can be taken and tonnage from a lower grade source may be added which under normal circumstances would not be available.

b. Whether ore grade is being concentrated, for example, by biological or chemical leaching. Here the residual enriched portion needs to be sampled effectively to obtain a representative grade for the tonnage submitted to plant.

The following stage which needs to be reconciled is the mill bins. The tonnages remaining in the bins are normally measured once a month and the tonnages and grades assigned are dealt with in the following way:

a. The tonnages at the beginning of the month that are in the mill bins are assigned the average grade of the last month. This in turn will be wholly depleted by the mill (unless, of course, the mill has a stoppage for the whole month)
b. Taking the above into consideration, the grade assigned to the mill bins at the end of the month will obviously be that grade sent to the mill during the month.

If one has a tight control on tonnage across the belt (a Go Belts sampler sampling at regular intervals) and knowledge of mill flow, one can determine tonnages in mill bins at end of month and calculate the grades of these from the belt samplings. This gives us a solid figure that can be relied upon unless there is theft at the plant of recovered ore. In contrast, determining a representative grade for residues is extremely difficult. This is generally caused by the following factors.

a. Irregular sampling intervals.
b. Obtaining a representative sample from the flow of slurry, which under most circumstances is under a certain degree of force. This in turn requires large and sophisticated sampling tools and methods to ensure a sufficient and representative sample.
c. Substituting the last value assayed if the flow has not been sampled on time, or sampling at inordinately high time intervals during which substantial changes in grade may occur and never be observed.
d. Not sampling at all and merely utilising a "historical" average. And yes, this happens!

From this process, a series of factors are obtained, such as the tonnage discrepancy, shaft call factor, the plant call factor, the belt call factor and the mine call factor to mention but a few. However, it is important to understand that these are not to be utilised as an excuse for poor recovery performance (e.g., "Yes, but the mine call factor is always around 88%"), but rather indicators as to where to investigate and take appropriate remedial action.

A mine-to-mill reconciliation must never be considered as a means of reconciling what has happened and then ignoring it; nevertheless, if a certain factor persistently occurs at a certain level and nobody does anything about it, this factor must be included in any estimate of available content to be mined or declared in a competent person's report.

Finally, if there is subsequent sampling or drilling during the reconciliation period, a block call factor can be calculated. This essentially entails updating the estimation block model with the latest data and then comparing the grades assigned to the areas mined before the information update to those obtained after the information update, thus supplying an indication of how additional information changes the value of the estimates. Most mines stop at the stage of calculating a single block factor. However, this need not be the case. If one obtains multiple months of information and, in addition, segregates these according to individual zones, much invaluable information for estimation purposes (e.g., extents of smoothing or conditional bias) can be obtained by plotting "the grades called for" to those estimated after additional info has been obtained.

FORMALISING THE RECONCILIATION FRAMEWORK

Due to variability in reconciliation processes, AMIRA formulated the Code of Practice for Metal Accounting and defined it as the process of estimating saleable metal quantities within a mine and its processing streams over a defined period (Gaylard et al., 2009). This involves comparing estimates from various sources during specific timeframes, a practice referred to as reconciliation (Morrison, 2008). Traditionally, metal accounting has focused on reconciling geological mineral resource estimates with mineral reserve conversions, reconciling long-term mine plans with short-term plans, matching head grades with sampled or block grades, and reconciling tonnage discrepancies and mine call factors through mine surveys.

Additionally, metallurgical reconciliation has relied on metal balancing, followed by commercial metal accounting linking dispatched products to sales quantities, quality, and revenues – a term known disparagingly as 'millmatics' among miners. However, these reconciliations have often been carried out independently for specific departmental purposes, lacking integration into an end-to-end process. To address this issue, AMIRA introduced the concept of an end-to-end metal accounting system, leading to the formation of the Metal Accounting Project P754 in 2003. The resulting Code for Metal Accounting was published in 2007, with a focus on mineral processing without fully incorporating geological and mining components due to limited support from mining companies.

The AMIRA Code principles and process can be extended to encompass the entire mining value chain as a comprehensive end-to-end reconciliation system (Gaylard et al., 2009), as shown in Figure 10.3. Corporate governance is to encompass all levels of a company, requiring integrated risk management and reporting throughout the organisation. Several factors endorse this approach, including Sarbanes-Oxley requirements (Seke, 2014), King III requirements for integrated reporting (Butler and Butler, 2010), the development of the ISO 31000 standard on risk management (Purdy, 2010), increasing shareholder demands for control and optimisation of products and earnings (Macfarlane, 2015), and the need for more transparent reporting of mineral resources and reserves (Butler and Butler, 2010; Gaylard et al., 2009; Macfarlane, 2015).

Adopting an end-to-end approach to metal accounting along the value chain is crucial for various reasons. It ensures a solid foundation for metal accounting, enables reconciliation of product mass at different points along the value chain, safeguards

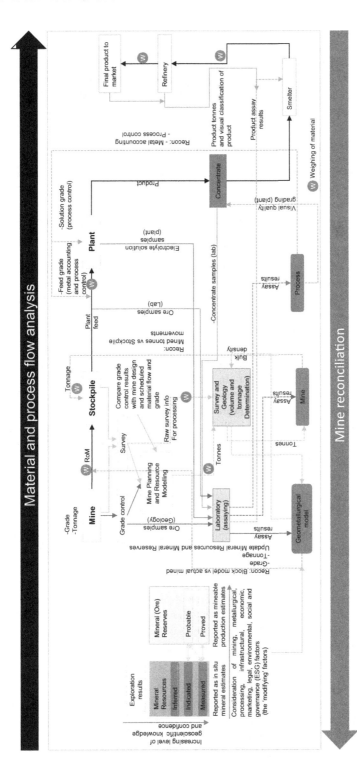

FIGURE 10.3 Critical components of end-to-end-reconciliation.

grades and products from losses, converts mineral resources effectively into saleable products, defines acceptable tolerances and ranges of variability, identifies and monitors modifying factors such as dilution and recovery, ensures competence and control in critical variables, and allows reliable reporting in public reports. Moreover, integrating metal accounting and reconciliation is important for efficient mineral asset management and enterprise risk management.

CHALLENGES IN MINE-TO-MILL RECONCILIATION

The mining environment presents unique challenges, requiring integrated reconciliation systems. These challenges include variations in base metal deposits, downstream processes as value-creating stages, different mining methods necessitating various grade control and tonnage reconciliation approaches, and critical dilution control to identify and manage its impact on net present value. Adopting a systematic approach is essential due to reconciliation's complexity and multidisciplinary nature. The AMIRA framework provides a suitable basis that can be adapted for such a system. The systematic process involves defining the purpose and objectives of the reconciliation exercise, determining reconciliation boundaries, and establishing a strategic plan system (Gaylard et al., 2009).

Various reconciliation arcs can then be identified along the value chain to address both management control and commercial reconciliation requirements. Another main challenge is the accuracy and reliability of the data. This can be affected by a range of factors, including errors in sampling or analysis, inconsistencies in data collection or reporting, and changes in the ore characteristics over time. Readers and practitioners also need to be aware of the complexity of the mining and processing operations, which can make it difficult to identify the causes of the differences between the actual and expected production data. Finally, there may be limitations in the tools and techniques available for data analysis and interpretation, which can affect the accuracy and effectiveness of the reconciliation process.

CONCLUSION

Mine-to-mill reconciliation is an important process in the mining industry, which can help to improve the efficiency and effectiveness of the mining and processing operations. By comparing the actual production data with the expected production data, and identifying the causes of any differences, mining companies can take corrective actions to improve the performance of the operation. While there are challenges associated with mine-to-mill reconciliation, these can be overcome through careful data collection and analysis, as well as ongoing monitoring and review of the mining and processing operations.

REFERENCES

Butler, R. & Butler, M. J. (2010). Beyond King III: Assigning accountability for IT governance in South African enterprises. *South African Journal of Business Management, 41*(3), 33–45.
Gaylard, P. G., Morrison, R. D., Randolph, N. G., Wortley, C. M. G., & Beck, R. D. (2009, July). Extending the application of the AMIRA P754 code of practice for metal accounting.

In Fouche, P.A.P. (Ed), *Proceedings of the 5th base metals conference* (pp. 15–38). Johannesburg: Southern African Institute of Mining and Metallurgy.

Macfarlane, A. S. (2015). Reconciliation along the mining value chain. *Journal of the Southern African Institute of Mining and Metallurgy*, 115(8), 679–685.

Morrison, R. D. (2008). *An introduction to metal balancing and reconciliation*. Australia: Julius Kruttschnitt Mineral Research Centre (pp. 1–618).

Purdy, G. (2010). ISO 31000: 2009—Setting a new standard for risk management. *Risk Analysis: An International Journal*, 30(6), 881–886.

Seke, D. (2014). From metal to money: The importance of reliable metallurgical accounting. *Journal of the Southern African Institute of Mining and Metallurgy*, 114(1), 1–6.

EXERCISE FOR CHAPTER 10

1. What is grade and tonnage reconciliation?

   Answer: Grade and tonnage reconciliation is the process of comparing the estimated tonnage and grade of an ore deposit, based on the geological model and resource estimation methods, with the actual production data from the mining operation.

2. Why is grade and tonnage reconciliation important?

   Answer: Grade and tonnage reconciliation is important for assessing the accuracy and reliability of the resource estimation methods and for identifying areas of potential improvement in the mining operation.

3. What are the key steps involved in grade and tonnage reconciliation?

   Answer: The key steps in grade and tonnage reconciliation include collecting and analysing the data, comparing the estimated and actual production data, identifying the causes of any differences, and implementing corrective actions if necessary.

4. What are some of the challenges associated with grade and tonnage reconciliation?

   Answer: Challenges associated with grade and tonnage reconciliation can include errors in sampling or analysis, inconsistencies in data collection or reporting, changes in the ore characteristics over time, and limitations in the tools and techniques available for data analysis and interpretation.

5. What are some of the tools and techniques used for grade and tonnage reconciliation?

   Answer: Tools and techniques used for grade and tonnage reconciliation can include statistical analysis, block modelling, geostatistical methods, and other data analysis tools.

6. What is the role of the geologist in grade and tonnage reconciliation?

   Answer: The geologist plays a key role in grade and tonnage reconciliation by providing input to the resource estimation process, monitoring the mining and production data, and interpreting the geology of the deposit.

7. How can grade and tonnage reconciliation be used to improve the mining operation?

   Answer: Grade and tonnage reconciliation can be used to identify areas of potential improvement in the mining operation, such as optimising the mining method, improving the accuracy of the resource estimation methods, and reducing waste.

8. What are some of the potential benefits of effective grade and tonnage reconciliation?

   Answer: Some of the potential benefits of effective grade and tonnage reconciliation can include increased resource efficiency, improved financial performance, and better management of risk.

9. What are some of the potential risks associated with ineffective grade and tonnage reconciliation?

   Answer: Risks associated with ineffective grade and tonnage reconciliation can include inaccuracies in resource estimates, inefficient mining operations, and reduced financial performance.

10. How can mining companies ensure effective grade and tonnage reconciliation?

    Answer: Mining companies can ensure effective grade and tonnage reconciliation by implementing rigorous data collection and analysis processes, using accurate and reliable resource estimation methods, and regularly reviewing and updating their geological models and mining plans.

11 Cut-off grade, net present value, and profit margin optimisation

INTRODUCTION

This chapter considers the effective utilisation of a cut-off and the implications for optimising Net Present Value (NPV) and profit margins. Cut-off grades are important in mining operations as they define the minimum amount of valuable product or metal that must be present in a metric tonne of material before it is sent for processing. They help distinguish the material to be processed from that which should be discarded. Additionally, cut-off grades play a significant role in deciding the routing of mined material between different processing methods, such as heap leaching and milling. They also influence whether the material should be processed immediately or stockpiled for future processing. Calculating cut-off grades involves comparing costs and benefits. In simple cases, a single value such as minimum metal content suffices to define the cut-off grade. However, in most situations, cut-off grades vary based on the material's geological characteristics, considering factors such as costs, recoveries, and environmental impacts. The grade is typically the most critical factor, but other considerations like acid-generating potential, refractory factor, and clay content may also influence the cut-off grade.

The cut-off grade not only determines the profitability of a mining operation but also affects the mine's lifespan. A higher cut-off grade can boost short-term profitability and net present value, benefiting shareholders and financial stakeholders. However, it may lead to a shorter mine life, limiting opportunities and beneficial socio-economic impacts. In contrast, lower cut-off grades can extend the mine life and maintain profitability during price downturns. External factors, such as metal prices and financial institution requirements, may also influence cut-off grade decisions. Conscious decisions can be made to increase mining capacity while keeping processing capacity constant, allowing for higher cut-off grades. Some lower-grade materials can be stockpiled for future processing, leading to various consequences like extended processing facility life and potential environmental risks. Cut-off grades significantly impact reserves subject to stock exchange regulations and accounting practices. Published reserves affect capital depreciation, book value, production cost, taxes, and market valuation of mining companies. Therefore, ensuring an accurate and unbiased estimation of reserves is essential to inform investors and stakeholders correctly.

Various stakeholders, both internal and external, have an interest in cut-off grades and the resulting reserves. Insiders include company management and employees, while outsiders encompass shareholders, financial institutions, local communities, environmentalists, regulators, and buyers. Cut-off grades should be calculated based on technical and economic constraints, but stakeholders' diverse interests and objectives need to be understood and prioritised to make informed decisions. The technical literature offers numerous publications on estimating and optimising cut-off grades, with the primary objective being to optimise the net present value of future cash flows. This optimisation involves considering spatial and temporal variables, the geographic location of the deposit, the order of material mining and processing, and the resulting cash flow. Mathematical solutions to cut-off grade optimisation can be complex due to the nature of the problem, incorporating both time and space considerations.

THE MECHANICS OF CUT-OFF GRADE

Cut-off grade is defined as the minimum amount of valuable product or metal that one metric tonne (1,000 kg) of material must contain before it is sent to the plant. Furthermore, it discriminates against material that should be discarded from that which should be processed (Rendu, 2009). According to Lane (1988), the value of a mine is derived from future projects' cash flows, while cash flows are based on three components of mining operations, namely mineralised material (development), ore (treatment), and mineral (marketing). Hence, there are mathematical formulae to determine three limiting cut-off grades and three balancing cut-off grades that will be determined to provide an optimum cut-off grade for each component. The research by Lane (1988) remains the standard mathematical formulation for solutions to cut-off grades when the objective is to maximise the NPV. Progress has been made to improve mine planning and optimise cut-off grades after Lane (1988). Complex algorithms and computer programs have been written to assist in analysing mine plans, testing the options, and improving production schedules.

The study by Minnitt (2004) re-examined the work by Lane (1988) regarding the optimisation of cut-off grades in mining operations. The work was based on calculus and the NPV criterion, which is mostly understood, consistent and appropriate for sequential cash flows arising from the extraction of ore. However, the method was found to be limiting because it considered NPV at various points of value chains to determine the cut-off grade. Table 11.1 is a summary of the factors that drive the cut-off grade with the description of how the cut-off grade is affected by the change of the factor.

PRINCIPLES OF CUT-OFF GRADE DETERMINATION

The principles of calculating and optimising cut-off grades can be found in mining and mineral processing literature (see Birch, 2017; Hall, 2014; Krige et al., 2004; Lane, 1988; Minnitt, 2004). Most of the uncertainties in cut-off grade calculation are caused by:

- Incorrect grade and tonnage estimation methods: some grade and tonnage estimation methods may overestimate, or underestimate volume extracted, density, and other factors affecting cut-off grade (Paithankar et al., 2020).

TABLE 11.1

Summary of the factors that drive the cut-off grade with the description of how the cut-off grade is affected by the change of the factor

Factors	Description
a. Metal price	When determining the optimum cut-off grade, the relationship between the cut-off grade and price is often the prominent determinant (Khodaiari and Jafarnejad, 2012). There is a direct proportionality between the metal price and the cut-off grade: that is, when the price of the metal, such as gold, increases, the cut-off grade also increases, and when the metal price decreases, the cut-off grade also decreases (Khodaiari and Jafarnejad, 2012) Notably, the lowest metal prices often lead to the highest metal grades.
b. Working cost (including ORD); processing cost and amp; mining cost tramming/ transport/dump cost	There is a positive relationship between costs, such as working costs, processing costs, mining costs (i.e. extraction costs), and cut-off grade (Krautkraemer, 1988). As such, when the costs increase, the cut-off grade also increases.
c. Volume mined and density.	The higher the volume and density mined or extracted, the higher the possibility of maximising the cut-off grade (Ahmadi, 2018).
d. Survey modifying factors – (PRF, MCF, shortfall, OSS, and development)	Survey modifying factors refer to the social, economic, and technical parameters that are used when converting raw mineral deposits into ore (profitable metals) (Birch, 2016). These are often limiting factors when calculating cut-off grades (Ahmadi, 2018).
e. Revenue (single or multi-metal recovered)	The higher the cut-off grade, the higher the revenue generated, as the relationship between cut-off grade and revenue is directly proportional (Osanloo and Ataei, 2003). Additionally, the cut-off grade is highly dependent on the difference between the costs incurred and the revenue generated (Osanloo and Ataei, 2003).

- Deposit drill hole and sampling data paucity: small deposit drill holes and data paucity or insufficient samples can result in inaccurate cut-off grade calculations (Courtney-Davies et al., 2019).
- High nugget effect and poor sampling practices: poor sampling practices can result in a high nugget effect, high variability amongst closely spaced samples, and thus, poor reliability, which makes the cut-off grade inaccurate (Guedes et al., 2020).

- Mineralogical different ore types: Different ore types during metallurgical recovery will cause inaccuracies in grade control hence directly affecting the cut-off grade (Rendu, 2009).
- Multiple process routes: Multiple steps in the determination of a cut-off grade might result in errors (Annels, 1991).
- Deleterious elements: these are elements found within other elements or deposits, such as gold and silver, and they result in errors in metal concentrations which in effect reduce the accuracy of the cut-off grade concentrations (Shirazi et al., 2021).
- Stockpile strategy: when stockpiled or bulk mining materials are used for volumetric calculation in the cut-off grade calculations, they introduce cumulative errors as some volumes are not processed, and so reduce the accuracy of cut-off grade calculations (Guedes et al., 2020).
- Blending constraints: some deposits or materials only blend a certain number of oils resulting in constrained volumes when calculating cut-off grades (Saliba and Dimitrakopoulos, 2019).
- By-product credits and metal equivalence: metal equivalence refers to mineralisation containing several metals of economic value and is converted to a single metal. This implies that minor metals are converted by formula and added to the grade of the major metal (Annels, 1991). Ignoring these uncertainties in daily mining operations may have very negative economic implications for a mining project due to the incorrect determination of the cut-off grade (Githiria and Musingwini, 2019).
- Net smelter returns: net smelter returns are revenues that are to be paid for metals mined and can reduce the accuracy of cut-off grades when they are not accounted for in the calculation (Wellmer et al., 2023).

Therefore, the methods and parameters used to calculate the cut-off grade must provide reasonable prospects of eventual economic extraction and should be aligned to the various stages of mineral resource development. For example, cut-off grades should be calculated differently during mineral resource exploration stages versus new versus mature mines. The generic cut-off grade calculation involves gauging the extraction costs and operating efficiencies using benchmarks from analogous mines and determining commodity prices from public reports or price forecast research notes. Cut-off grade is usually expressed in grade units that may require conversion from market pricing units.

For the cut-off grade, the price needs to be expressed in grams, for example:

$$\text{Cut-off grade} = \text{Operating cost}\,(\text{US\$})\,\text{metric tonne}\,(\text{t}) \times \text{Grade based revenue} =$$
$$60\,\text{US\$t} \times g\,34.7\,\text{US\$} = 1.73\,\text{g/t}$$

COMMON METHODS OF CUT-OFF-GRADE DETERMINATION

The cut-off grade is a set of criteria used in mining operations to distinguish ore from the waste within a mineral deposit (Bascetin and Nieto, 2007). Waste can be left where it is or transported to landfills, while ore is transported to a treatment plant for further processing (Bascetin and Nieto, 2007). It delineates a portion of mineral resources or mineral reserves that has an average grade designed to meet specific goals or results (The AusIMM Bulletin, 2016). There are three ways of determining a cut-off grade which are the break-even method, the heuristic method, and Lane's optimum cut-off grade method (Githiria and Musingwini, 2019). However, this chapter only focuses on two methods, namely the break-even method, which considers the economic parameters only, and Lane's method, which relies on the optimisation of the net present value.

a. The break-even method

The break-even method is defined as the grade whereby the revenue generated equals the cost of producing that revenue. A break-even cut-off grade model is a one-dimensional process since it only takes into consideration the economic characteristics and overlooks both the deposit's grade-tonnage distribution and production capacity (mine, processing, and refining capacities) (Githiria, 2016). It is mostly used by mining companies for the estimation of ore reserves and planning since it is simple. The formula for determining the break-even grade is:

$$\text{Breakeven grade (units / t)} = \frac{\text{costs (\$ / t)}}{\text{payability (\%)} \times \text{product price [\$/units]} \times \text{recovery [\%]}}$$

The disadvantage of the break-even formula and model is that the formula cannot be used to achieve the aims of the mine, such as to deliver shareholder value (The AusIMM Bulletin, 2016), and the model assumes that every tonne classified as ore pays for itself at the time it is treated (Githiria, 2016). As a result, the use of a break-even cut-off grade will lead to mine plans that are almost guaranteed to not deliver the company's stated goals (The AusIMM Bulletin, 2016).

b. Lane method

The cut-off grade is an operational planning parameter that is defined as the minimum profitable mining panel (or block) grade (e.g. g/t or vol.%) and is determined by a suitable metric for extraction under a set of economic and operational circumstances (Collard, 2013; Lane, 1988). The definition mentioned above most certainly leaves one with a distinct 'Huh, what was that?' feeling. Let us try and explain this concept in a way which is simple and yet comprehensive enough to ensure a feeling of complete understanding in the reader. Firstly, one needs to point out that there is more than one cut-off form. One can have a cut-off which is derived from the economic break-even. As an aside, the economic break-even is that average

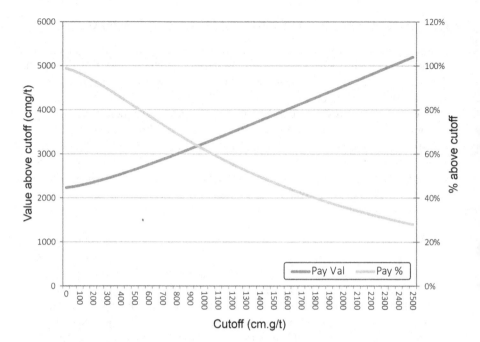

FIGURE 11.1 Example of a grade-tonnage curve.

value mined that will result in a financial break even for the mine, which is neither a profit nor a loss. These calculations are dealt with elsewhere in the document. An alternative cut-off is that calculated from the economic break-even value plus a margin added to that economic break-even. The first is normally referred to as the economic break-even cut-off, and the second is the margin cut-off.

Now, let's provide a definition that explains what the cut-off really is. If the cut-off is based on the economic break-even this is the lowest value that can be mined, such that, if the planning and execution of the plan is done in a diligent fashion (i.e. we will get to what this means a little later), the average value mined will be equal to the economic break-even. A similar situation applies to the margin cut-off here, the resultant average value mined will be equal to the economic break-even plus the selected margin. The question now arises as to how does one calculate these values? A simple and intuitive method is the grade-tonnage curve route. Here, one must realise that the standard deviation used for the grade-tonnage curve calculation must be that representing the selective mining unit (SMU). This will allow for cut-offs to be calculated that are representative of the relevant mining selectivity. The process requires that for each domain that mining will be taking place in, one calculates the corresponding grade-tonnage curve, as shown in Figure 11.1.

One notices that here with the traditional grade-tonnage curve, the cut-offs are on the X-axis and the average grade above cut-off is on the Y-axis. We want to obtain the

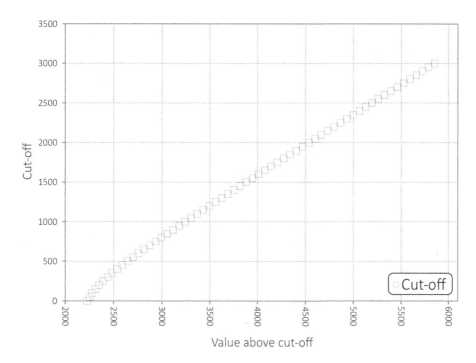

FIGURE 11.2   Example of grade-tonnage curve with axes switched.

cut-off for an average grade above the cut-off. Hence, we will switch the axis, making the value above cut-off the independent variable on the X-axis and the cut-off the dependent variable on the Y-axis (Figure 11.2).

Now, supposing one requires the cut-off for an economic break-even of 2,500, one can interpret the position and say that the cut-off is at about 400. Alternatively, one can fit a trend line to the data, obtain the equation, and plug in the value of 2,500 (Figure 11.3).

Here by plugging the figure of 2,500 into the equation provided on the chart, a cut-off of 340 is obtained. For example, let's say a 10% margin on this economic break-even utilising 2,750, one obtains a margin cut-off of 576.

From the above example, one observes that the calculation of individual cut-offs is simple. However, on an actual mine, the area mined will be across multiple geostatistical domains, where the grade-tonnage curves are definitely not the same. The starting point to overcome this problem is to calculate the grade-tonnage curves for each geostatistical domain. However, this is just the starting point; from here, there are two methods that can be utilised, the first being one calculates a weighted average cut-off for the mine based on the amount expected to be mined from each domain (Table 11.2).

This method has the advantage of being simple to calculate and to implement, with the whole mine having a single cut-off. Therefore, there is not a single domain that is utilising the correct cut-off. This, in turn, could have disastrous implications should the mining in a domain not result in the expected results. Consider if domain 2 which has a cut-off of 452 had to exceed call and mine 8,000 sqm and domain 5 had to have a fall of ground and mine nothing (Table 11.3).

TABLE 11.2

Example of weighted mean cut-off calculation

Domain	Area Planned (m²)	COG (Au (cm.g/t))	Area Planned (m²) × COG (Au (cm.g/t))
1	3,500	386	1,351,000
2	5,000	452	2,260,000
3	2,650	197	522,050
4	8,000	410	3,280,000
5	1,500	122	183,000
Total	20,650		7,596,050
Average COG		368	

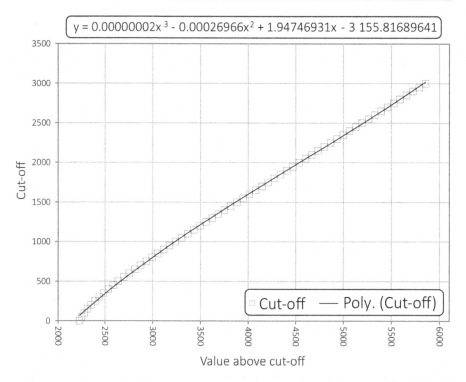

FIGURE 11.3 Example of polynomial trend function fitted to the chart.

The resultant cut-off required for the mine should now have been 396, a difference of 28 cm.g/t. Working this back to the average values mined results in an 8% loss for the mine. Not a very desirable situation! Thus, one needs to ensure that calls are strictly adhered to if one wants to use this method; however, as practically this very rarely if ever occurs the method described should only be considered as a last resort or if only a single domain is being mined. Notwithstanding what was said before in the case of an opencast mining situation where under most circumstances, everything has to

TABLE 11.3

Example of weighted mean cut-off of subsequent actual mining

Domain	Area Planned (m²)	COG (Au (cm.g/t))	Area Planned (m²) × COG (Au (cm.g/t))
1	3,500	386	1,351,000
2	8,000	452	3,616,000
3	2,650	197	522,050
4	8,000	410	3,280,000
5	0	122	0
Total	22,150		8,769,050
Average COG		396	

be blasted; thus, cut-offs are only applied post blasting and pertain to where the tonnage needs to be trammed to, that is, for example, either to the low-grade stockpile or to the mill. Under these circumstances, selection of what goes to the mill can be obtained post blasting. The second method requires more upfront work for the person providing the cut-offs and the planning officer to take cognisance of where he/she is planning in order to ensure the correct cut-off is utilised. Here the person providing the cut-offs needs to provide these in such a manner that it is relatively easy for the planner to determine, while planning, as to what cut-off applies to what area. This can be accomplished relatively simply if the planner is planning in a graphical environment and is able to display the domains on his computer screen while planning. All that is required in addition to this is that a table be available to the planner with domains and corresponding cut-offs. A quick glance upward at the wall or the second screen provides the relevant cut-off.

At this stage, the authors, unfortunately, have to burst the readers' bubble and point out that there is an additional layer of work that needs to be done in order to ensure that the correct grade is mined, and that the planner does not waste time planning and re-planning multiple times in order to get the grade planned correct. Too often, the mindset obtained from utilising a cut-off is that one now has carte blanche to mine as much as they want in any grade range as long as it is above the cut-off grade. Nothing could be further from the truth! There are two factors that militate against this:

The first being that, due to high grading in the past, an underground mine normally has huge amounts of lower-grade ore available to mine and much less high-grade. This ore may often fall within the range above the cut-off and thus result in a disproportionate amount of ore being mined within this range. Consider the extreme example of a period in which all the ore mined fell within the range of above cut-off but below economic break-even. Under this circumstance, it would be impossible to obtain a profit for that period and the mine would be operating at a loss. From the aforementioned, it should be now evident that one cannot only utilise the criteria above cut-off; rather, this should be a combination of grade ranges and proportions that can be mined and strictly adhered to for each grade range. See some examples showing the differing proportions that are available to mine and need to be mined in these proportions in order to ensure the required grade to mill (Figure 11.4(A) and 11.4(B)).

Let us now consider the net present value and how this can be optimised by utilising the distribution of values within the orebody to obtain the optimal cut-off that maximises the NPV. This is almost certainly not the highest margin, and is an erroneous opinion held by many within the mining industry.

The process to determine the optimal cut-off that simultaneously optimises the NPV is as follows:

1. Determine the grade-tonnage curve parameters for the mine or the portion under consideration. This needs to consider the parameters below:
   a. The SMU for the area under consideration.
   b. Whether the mineral under consideration is of a normal, two-parameter log-normal or three-parameter log-normal distribution. This will impact how the dispersion variance is calculated.
   c. The dispersion variance of the SMU mentioned in a. How this is calculated is explained under section 7.8 Calculating the dispersion variances.
2. Determine the mining rate for the area under consideration. For example, 5,000 m² per month. Liaise with the planner for possible increases or decreases in volumes mined across the periods under review. Remember this will under most circumstances, result in a different economic break-even and hence impact on the process for point 4.
3. Calculate the grade and tonnage curve utilising multiple cut-offs.
4. For each cut-off that provides a grade above cut-off that is above the economic break-even:
   a. Utilising the mining rate determines the period available to mine.
   b. The volume of metal recovered. Parameters required for this calculation are mentioned in the sections on mine-to-mill reconciliation and economic break-even calculations.
   c. The future value of the recovered mineral under consideration for each period mentioned in 4a.
   d. The future value of cost of mining. Escalation rates for the mineral and the cost of mining may differ in this respect.
5. Determine the discount rate to be applied to the future values of revenues obtained. Some considerations will need to be:
   a. The number of periods available to mine.
   b. Local and international situations that may affect the revenue and costs involved in mining.
   c. Is the period considered short-, medium-, or long-term?
6. Subtract the future costs from the revenue achieved for each cut-off scenario and each period. This will provide future profit for each cut-off scenario and period.
7. Apply the discount rate to each scenario and period depending upon the period from start of the project.
8. Summate the discounted periods for each scenario. This will provide the NPV for each scenario.

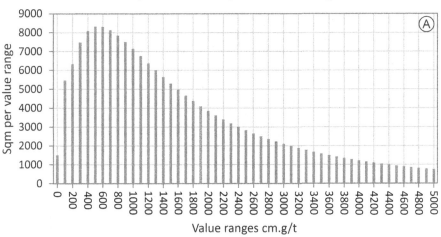

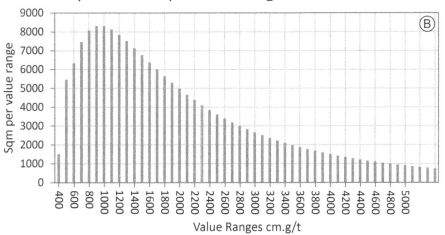

FIGURE 11.4 (A) Ore grade range per square metre, and (B) Ore grade range per square metre.

9. Plot the scenarios with positive NPVs on a graph with the cut-offs on the X-axis and the NPVs on the Y-axis.
10. Determine by examination the optimal NPV.
11. For the optimal NPV, determine the margin that needs to be planned. This is obtained by the formula (revenue/costs) – 1.

All of the above can be accomplished simply within a spreadsheet with economic break-evens and volumes per period being available in the first two rows of the spreadsheet. Subsequent formulas can then take cognisance of these parameters.

Most of the uncertainties in cut-off grade calculation are caused by:

a. Incorrect grade and tonnage estimation methods: Utilise the section and parameters on grade-tonnage calculations.
b. Deposit drill hole and sampling data paucity: consider the methods to obtain a representative variance from minimal sampling data.
c. High nugget effect and poor sampling practices: Stick to the sampling practices that provide a representative sample outlined in this document.
d. Mineralogically different ore types: Ensure correct domaining practices as outlined in this document; thumb sucks must be avoided at all costs.
e. Multiple process routes: Have a peer check your calculations explaining your rationale. Two heads are better than one, and three heads better than two.
f. Deleterious elements: Ensure that drill splits are done regularly and sufficient volume to ensure metal content is represented.
g. When utilising the orebody as the constraining mechanism the impact of stockpile strategies is minimised as all of the ore will eventually be milled. Creating low-grade stockpiles that wait for a beneficial increase in metal prices will only increase the NPV.
h. Blending constraints: This is a concern if the plant does not ensure proper management of ore blending or if stockpile management is not accomplished rigorously.
i. By-product credits and metal equivalence: metal equivalence refers to mineralisation containing several metals of economic value and is converted to a single metal. This method is of a deleterious nature; it is far better that the necessary assay methods be employed (albeit at a slightly higher assay cost and a slightly larger workload for the evaluator) in order to ensure that the elements may be estimated individually.
j. Metal prices. Here one needs to review and recalculate (i.e. not a big deal if a process template has been set up) NPV optimised cut-offs when there is a significant change in price (Table 11.4).

TABLE 11.4
Factors included in the calculation of the NPV

Variable	Description
Mean grade	The mean value of the gold grade of panels available for mining over the entire subject mine's mining lease area.
Variance	Dispersion variance of SMU as deduced from the mine dataset.
Beta (β)	This is a shift parameter that allows for adjustment of the transformed data. The addition of β provides an additional degree of freedom to adjust the transformation, potentially improving the fit to a normal distribution. The value of β should be chosen carefully. It is typically a small positive number added to shift the distribution appropriately. It may be determined empirically or through optimization methods.
Discount rate (α)	Discount rates commonly used within the mining industry range between 5% and 15%.
Price of mineral or metal	Based on commodity price and escalation rate if required.
Time period (n)	For NPV calculation over the calculated time periods.

CONCLUSION

To evaluate a mining project's economic feasibility, it is crucial to determine the cut-off grade because it influences the tonnage and grade of the processed material. A mine's cut-off grade profile determines the mine's size, processing plant capacity, and free cash flow through an iterative process. In making this decision, all consequences must be considered, including technical, economic, legal, environmental, social, and political factors that may be affected. The significance of cut-off grades extends to various aspects of the mining operation, such as defining tonnages mined and processed, average grade of mill feed, cash flows, mine lifespan, and other major characteristics. However, the impacts of cut-off grade changes go beyond financially quantifiable effects. There are other costs and benefits associated with these changes, even if they are not easily measurable. Given the interrelated nature of cut-off grade with mining and processing capacities, costs, market value of products, and cash flow, it is essential to thoroughly review all opportunity costs and potential benefits before implementing a change in the cut-off grade. Declining cut-off grades may maximise net present value but could lead to reduced total undiscounted revenues from sales. Conversely, increasing the cut-off grade might result in discarding profitable low-grade material. In such cases, it might be worth considering the option of stockpiling lower-grade material for future processing. The dynamic nature of mining projects necessitates continuous re-evaluation of previously estimated optimal cut-off grades. Changes in costs, prices, and mine and mill performance can significantly impact future cash flow and opportunity costs. While maximising net present value is important, it is crucial to avoid disregarding actions that may only bear consequences towards the end of the mine's lifespan. To optimise cut-off grades while accounting for unquantifiable costs and benefits, one approach involves evaluating the project under various constraints, including discount rates, mine or mill capacity, sales volume, capital or operating costs, and other relevant factors. This comprehensive assessment can aid in making informed decisions regarding the cut-off grade, ensuring the overall profitability and sustainability of the mining operation.

REFERENCES

Ahmadi, M. R. (2018). Cutoff grade optimization based on maximizing net present value using a computer model. *Journal of Sustainable Mining*, *17*(2), 68–75.

Annels, A. E. (1991). *Mineral deposit evaluation, a practical approach*. London: Chapman and Hall, p 436.

The AusIMM Bulletin (2016). *Break-even is broken: Learn how to align your break-even grades with company goals*. AMC Consultants.

Bascetin A. & Nieto, A. (2007). Determination of optimal cut-off grade policy to optimize NPV using a new approach with optimization factor. *Journal of the Southern African Institute of Mining and Metallurgy*, *107*(2), 87–94.

Birch, C. (2016). Impact of discount rates on cut-off grades for narrow tabular gold deposits. *Journal of the Southern African Institute of Mining and Metallurgy*, *116*(2), 115–122.

Birch, C. (2017). Optimization of cut-off grades considering grade uncertainty in narrow, tabular gold deposits. *Journal of the Southern African Institute of Mining and Metallurgy*, *117*(2), 149–156.

Collard, J. (2013). Planification stratégique d'une mine souterraine avec teneur de coupure variable. Ph.D. Thesis, École Polytechnique de Montréal.

Courtney-Davies, L., Ciobanu, C. L., Verdugo-Ihl, M. R., Dmitrijeva, M., Cook, N. J., Ehrig, K., & Wade, B. P. (2019). Hematite geochemistry and geochronology resolve genetic and temporal links among iron-oxide copper gold systems, Olympic Dam district, South Australia. *Precambrian Research*, *335*, 105480.

Githiria, J. (2016). Cut-off grade optimisation to maximise the net present value using whittle. *International Journal of Mining and Mineral Engineering*, *7*(4), 313–327.

Githiria, J. & Musingwini, C. (2019). A stochastic cut-off grade optimization model to incorporate uncertainty for improved project value. *Journal of the Southern African Institute of Mining and Metallurgy*, *119*(3), 217–228.

Guedes, R. S., Ramos, S. J., Gastauer, M., Júnior, C. F. C., Martins, G. C., da Rocha Nascimento Júnior, W.,…., & Siqueira, J. O. (2021). Challenges and potential approaches for soil recovery in iron open pit mines and waste piles. *Environmental Earth Sciences*, *80*(18), 640.

Hall, B. (2014). *Cut-off grades and optimising the strategic mine plan*. Australasian Institute of Mining and Metallurgy.

Khodaiari, A. A. & Jafarnejad, A. (2012). The effect of price changes on optimum cut-off grade of different open-pit mines. *Journal of Mining and Environment*, *3*(1), 61–68.

Krautkraemer, J. A. (1988). The cut-off grade and the theory of extraction. *Canadian Journal of Economics*, *21*(1), 146–160.

Krige, D. G., Assibey-Bonsu, W., & Tolmay, L. C. K. (2004). Post processing of SK estimators and simulations for assessment of recoverable resources and reserves for South African gold mines. In Leuangthong, O., Deutsch, C.V. (eds), *Geostatistics Banff 2004. Quantitative Geology and Geostatistics* (vol. 14). Dordrecht: Springer. https://doi.org/10.1007/978-1-4020-3610-1_38.

Lane, K. F. (1988). *The economic definition of ore, cut-off grades in theory and practice* (1st ed.). London: Mining Journal Books Limited.

Minnitt, R. C. A. (2004). Cut-off grade determination for the maximum value of a small Wits-type gold mining operation. *Journal of the Southern African Institute of Mining and Metallurgy*, *104*(5), 277–283.

Osanloo, M. & Ataei, M. (2003). Using equivalent grade factors to find the optimum cut-off grades of multiple metal deposits. *Minerals Engineering*, *16*(8), 771–776.

Paithankar, A., Chatterjee, S., Goodfellow, R., & Asad, M. W. A. (2020). Simultaneous stochastic optimization of production sequence and dynamic cut-off grades in an open pit mining operation. *Resources Policy*, *66*, 101634.

Rendu, J. (2009). Cut-off grade estimation-old principles revisited-application to optimisation of net present value and internal rate of return. Orebody Modelling and Strategic Mine Planning conference, Perth.

Saliba, Z. & Dimitrakopoulos, R. (2019). Simultaneous stochastic optimization of an open pit gold mining complex with supply and market uncertainty. *Mining Technology*, *128*(4), 216–229.

Shirazi, A., Shirazy, A., Nazerian, H., & Khakmardan, S. (2021). Geochemical and behavioral modeling of phosphorus and sulfur as deleterious elements of iron ore to be used in geometallurgical studies, sheytoor iron ore, Iran. *Open Journal of Geology*, *11*(11), 596–620.

Wellmer, F. W., Scholz, R. W., & Bastian, D. (2023). Can ultimate recoverable resources (URRs) be assessed? Does analyzing declining ore grades help? *Mineral Economics*, *36*, 599–613. https://doi.org/10.1007/s13563-023-00368-0.

EXERCISE FOR CHAPTER 11

1. What is a cut-off grade?

    Answer: A cut-off grade is the lowest grade of mineralised material that can be economically mined and processed at a particular time.

2. What factors influence the determination of a cut-off grade?

    Answer: Factors that influence the determination of a cut-off grade include the commodity price, mining costs, processing costs, recovery rates, and other economic factors.

3. How is the cut-off grade calculated?

    Answer: The cut-off grade is calculated by comparing the revenue generated by mining and processing the mineralised material to the cost of mining and processing that material.

4. What is the role of the geologist in determining the cut-off grade?

    Answer: The geologist plays a key role in determining the cut-off grade by providing information about the grade and tonnage of the deposit, the mineralogy and metallurgy of the ore, and the geology and structure of the deposit.

5. How does the cut-off grade affect the mining operation?

    Answer: The cut-off grade can affect the mining operation by determining the amount of ore that is mined and processed, the profitability of the operation, and the overall efficiency of the mining process.

6. How can the cut-off grade be optimised?

    Answer: The cut-off grade can be optimised by adjusting the economic and technical parameters used in the calculation, such as the commodity price, mining costs, processing costs, and recovery rates.

7. What are some of the risks associated with using a cut-off grade?

    Answer: Risks associated with using a cut-off grade can include inaccurate resource estimates, inefficient mining operations, and missed opportunities for additional exploration and development.

8. How can mining companies ensure that the cut-off grade is appropriate for the deposit?

    Answer: Mining companies can ensure that the cut-off grade is appropriate for the deposit by using accurate and reliable resource estimation methods, regularly reviewing and updating the geological model, and incorporating new information as it becomes available.

9. What is the impact of changing commodity prices on the cut-off grade?

    Answer: Changing commodity prices can have a significant impact on the cut-off grade, as higher prices may make lower-grade mineralised material economically viable to mine and process, while lower prices may require a higher cut-off grade.

10. What are some of the potential benefits of using an optimised cut-off grade?

    Answer: Some of the potential benefits of using an optimised cut-off grade can include increased resource efficiency, improved financial performance, and better management of risk.

11. A copper deposit has an average grade of 0.8% Cu and a total production cost of $30 per tonne of ore. The current copper price is $7,000 per metric tonne. What is the cut-off grade for the deposit?

Answer: The cut-off grade can be calculated using the following formula:

$$\text{Cut-Off Grade} = \frac{\left(\text{Total Production Cost}\right)}{\left(\text{Copper Price} \times \text{Copper Recovery Rate} \times \text{Ore Grade}\right)}$$

Assuming a recovery rate of 80%, the cut-off grade is:

$$\text{Cut-Off Grade} = \frac{(\$30)}{(\$7{,}000/\text{tonne} \times 0.8 \times 0.008)} = 0.54\% \, \text{Cu}$$

12. A gold deposit has an average grade of 2.5 g/t Au and a total production cost of $50 per tonne of ore. The current gold price is $50 per gram. What is the cut-off grade for the deposit?

Answer: The cut-off grade can be calculated using the following formula:

$$\text{Cut-Off Grade} = \frac{\left(\text{Total Production Cost}\right)}{\left(\text{Gold Price} \times \text{Gold Recovery Rate} \times \text{Ore Grade}\right)}$$

Assuming a recovery rate of 90%, the cut-off grade is:

$$\text{Cut-Off Grade} = \frac{(\$50)}{(\$50/g \times 0.9 \times 2.5)} = \frac{0.444g}{tAu}$$

13. A zinc deposit has an average grade of 6.5% Zn and a total production cost of $40 per tonne of ore. The current zinc price is $3,000 per metric tonne. What is the cut-off grade for the deposit?

Answer: The cut-off grade can be calculated using the following formula:

$$\text{Cut-Off Grade} = \frac{\left(\text{Total Production Cost}\right)}{\left(\text{Zinc Price} \times \text{Zinc Recovery Rate} \times \text{Ore Grade}\right)}$$

Assuming a recovery rate of 85%, the cut-off grade is:

$$\text{Cut-Off Grade} = \frac{(\$40)}{(\$3{,}000/\text{tonne} \times 0.85 \times 0.065)} = 6.06\% \, \text{Zn}$$

14. A nickel deposit has an average grade of 2.2% Ni and a total production cost of $60 per tonne of ore. The current nickel price is $15,000 per metric tonne. What is the cut-off grade for the deposit?

Answer: The cut-off grade can be calculated using the following formula:

$$\text{Cut-Off Grade} = \frac{\left(\text{Total Production Cost}\right)}{\left(\text{Nickel Price} \times \text{Nickel Recovery Rate} \times \text{Ore Grade}\right)}$$

Assuming a recovery rate of 70%, the cut-off grade is:

$$\text{Cut-Off Grade} = \frac{\left(\$60\right)}{\left(\$15,000 \,/\, \text{tonne} \times 0.7 \times 0.022\right)} = 2.06\% \,\text{Ni}$$

15. A silver deposit has an average grade of 500 g/t Ag and a total production cost of $70 per tonne of ore. The current silver price is $750 per kilogram. What is the cut-off grade for the deposit?

    Answer: The cut-off grade can be calculated using the following formula:

$$\text{Cut-Off Grade} = \frac{\left(\text{Total Production Cost}\right)}{\left(\text{Silver Price} \times \text{Silver Recovery Rate} \times \text{Ore Grade}\right)}$$

Assuming a recovery rate of 80%, the cut-off grade is:

$$\text{Cut-Off Grade} = \frac{\left(\$70\right)}{\left(\$750 \,/\, \text{kg} \times 0.8 \times 0.5\right)} = 0.186 \,\text{g/t Ag}$$

Appendix 1
Log transformation

Log-normal transformation allows positively skewed data to be standardised to normal distribution for parametric statistics and semi-variogram modelling. In some cases, the back transformation may have negative implications.

THE 2-PARAMETER LOG-NORMAL DISTRIBUTION

A positively skewed variable x is log-normally distributed with 2-parameters μ and σ^2 if $Y = \ln(x)$ is normally distributed with mean μ and variance σ^2. Here, it is a simple manner to determine whether a distribution is 2-parameter log-normal. Simply take the natural logarithm of each value and determine if the resultant distribution satisfies a normal distribution function.

Notice the extended tail of the distribution an indication of a log-normal or 3-parameter log-normal distribution. Methods of determining whether the transformation is effective. This applies as a check for both 2-parameter and 3-parameter log-normal.

- Does the skewness approach zero? Anything with a skewness of less than 0.1 can be considered a reasonable fit. A point to take into consideration is although most software nowadays, and that includes Microsoft Excel, depicting a normal distribution as skewness of zero; however, one may encounter software that does not normalise the skewness parameter to zero, in which case one will have a value of three instead.

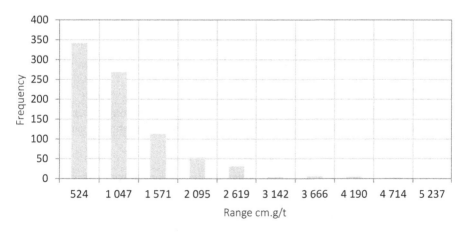

FIGURE A.1.1   Example of log-normal or 3-parameter log-normal distribution.

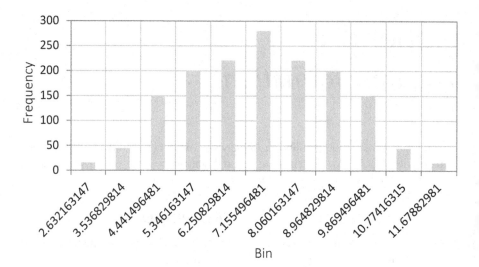

FIGURE A.1.2   Example of an effective 2-parameter transformation.

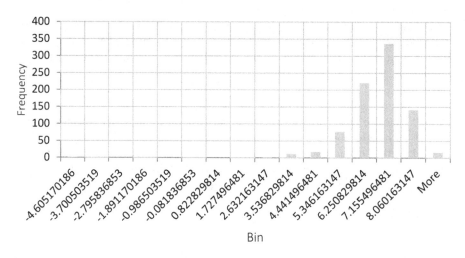

FIGURE A.1.3   Example of potential 3-parameter transform.

- A probability plot will produce a straight line. One may observe some
  scatter around the line; however, it is the overall shape that concerns only
  here.

Notice that the histogram has now become negatively skewed which is an indication
of the need for a 3-parameter transform. Additionally, the skewness of the dataset is
-2.77 another indication of requiring a non-zero third parameter.

Notice the distinct downward turn at the beginning of the plot here, once again, an
indication that the data requires a third parameter.

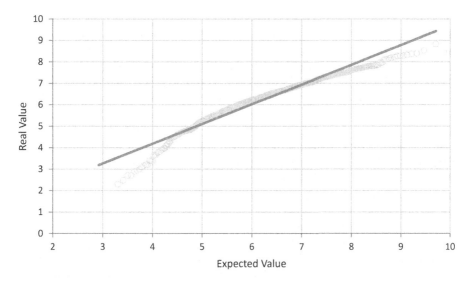

FIGURE A.1.4 Example of probability plot of histogram.

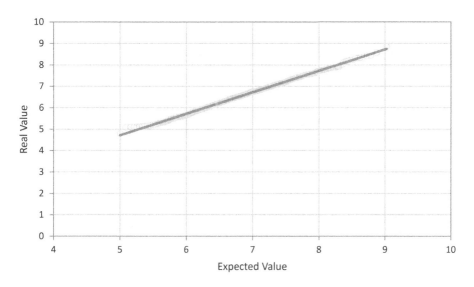

FIGURE A.1.5 Example of probability plot of the correctly fitted third parameter.

THE 3-PARAMETER LOG-NORMAL DISTRIBUTION

- A positively skewed variable x is log-normally distributed with 3-parameters μ, σ^2, and β if $Y = \ln(x + \beta)$ is normally distributed with mean μ and variance σ^2.
- β is often called the threshold parameter
- Back transformation: $= \mathrm{Exp}(\mathrm{Log}\ \mu + (\mathrm{Log}\ \sigma^2\ /2)) - \beta$

Here, the skewness is 0.0017, most certainly a decent fit. So, how does one obtain a decent fit of a third parameter in the case of a third parameter log-normal distribution? Although there are various methods put forward, some better some worse, in the author's opinion, the simplest and most effective method is that of minimising the skewness function. This can be done very simply by placing the dataset in an excel spreadsheet and adding an additional column which has a formula that adds the value, and a constant which is set in a cell to any arbitrary value for all the data. Only then takes the natural logarithm of the data plus the constant. Now, in an additional cell, one places the formula for the skewness of the data plus the third parameter. One then utilises the goal to set the skewness value to zero by changing the value of the third parameter. This will supply the lowest skewness possible by adding a third parameter. If one does not have the benefit of Microsoft Excel, one merely utilises the process of halves. Here, one calculates the skewness using an arbitrary third parameter (suggest 15% of mean of data) and only then considers the skewness; if it is still negative, one increases the third parameter; if not, one then halves the third parameter and examines the skewness. If the skewness is now negative, one adds half of the differences between the previous to the current value; if not, one halves the value once again. One continues either halving the value or adding half of the value difference until one achieves an acceptable skewness. This normally tends to quickly converge after four to five iterations.

Appendix 2
Normal score transformation

Normal score transformation is a powerful tool for understanding data. It reduces numerical artefacts that can obscure relationships while simplifying the portions of numerical modelling.

1. Compile the available data. Treat outliers as per section, thereon, ensuring before examining outliers that one is utilising data from generally speaking homogenous domains, just lumping all data together and then doing a normal score transformation may appear to be perfectly homogenous and perfect Gaussian distribution. After all, this is what a normal score transform does. However, the underlying inhomogeneity is merely being hidden and has not been effectively dealt with. If one requires the data to be declustered, decluster, and debias. If declustering weights are available, this is normally due to differing numbers of samples within regularised blocks and is the result of the regularisation calc is that only needs to ensure the same sort of practice utilised in stratified sampling, add weights to the percentile calculation.
2. Sort data, use the $n + 1$ percentile calculation that assumes tails (minimum and maximum) are not sampled. Assign minimum and maximum values.

     Also, calculate the associated Standard Normal, $N[0,1]$, percentile values using the Norm.Inv() (Microsoft Excel function).
3. Plot the Au (g/t) CDF.
4. Plot the N[Au (g/t)] CDF.
5. Graphically observe the transform of a single value in the table, $y = GY^{-1}F_Z(z)$.
6. Compare results to Percentile.EXC (Microsoft Excel function) that also uses a $N + 1$ basis but without explicit tail assignment (Error if outside data range).
7. Plot the $Q–Q$ plot to observe the Au $(g/t) – N$[Au (g/t)] transform function directly.

Back transformation is accomplished in the most effective manner by utilising a Hermite polynomial to the original dataset of the values before and after transformation. This ensures that the overall shape of the conversion function is retained, and linear interpolation is not required between values for values that were not in the original dataset.

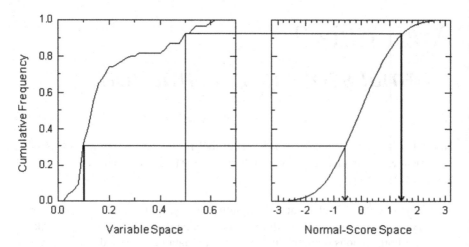

FIGURE A.2.1   Illustration of Normal Score data transformation.

Appendix 3
Q–Q plot (quantile–quantile plot)

A Q–Q plot (quantile–quantile plot) is a probability plot, which is a graphical method for comparing two probability distributions by plotting their quantiles against each other . First, the set of intervals for the quantiles is chosen. A point (x, y) on the plot corresponds to one of the quantiles of the second distribution (y-coordinate) plotted against the same quantile of the first distribution (x-coordinate). By a quantile, we mean the fraction (or percent) of points below the given value. That is, the 0.3 (or 30%) quantile is the point at which 30% percent of the data fall below and 70% fall above that value.

HOW TO CALCULATE A Q–Q PLOT?

The Q–Q stands for quantile–quantile which essentially means that one compares the expected quantiles (that is the standard deviations of the percentage points of the original dataset sorted ascending order) to the actual dataset also sorted in ascending order. This is quite a mouthful; however, it will become clearer as we progress through the process.

THE PROCESS

1. Get the data for the plot.
2. Now sort the data order ascending.
3. Number the data from one to the end of the data in an order ascending manner making sure not to miss and numbers, for example, 1, 2, 3, 4, 5, 6, etc.
4. Now divide each number generated in 3 by the total number of data plus one. The reason one does this is to ensure that one does not have the final percentage calculated equal to 1.
5. Now, for each percentage, calculate Inverse of the standard normal cumulative distribution function. NORM.S.INV (Microsoft Excel function).
6. Now plot an X, Y graph with the original values on the X axis and the values obtained in 5 on the Y axis.

The value pairs should be in approximately straight line. The size of the dataset governs the straightness of the line (e.g. larger datasets tend to produce better Q–Q plots if the underlying distribution is normal) and whether the samples are obtained in an unbiased manner.

EXAMPLE OF CALCULATION

TABLE A.3.1
Example of calculation of Q–Q plot

Step	Description	Excel Formula Example	Data Example
1	Input Data	---	5, 20, 15, 10, 25, 30
2	Sort Data	=SORT(A2:A7)	5, 10, 15, 20, 25, 30
3	Data Quantiles	=PERCENTILE.INC(A2:A7, K)	10th%, 20th%, ..., 90th%
4	Theoretical Quantiles	=NORM.INV((ROW(A1)-0.5)/ COUNT(A:A), MEAN(A:A), STDEV.P(A:A))	Corresponding normal distribution values
5	Plot Data	Insert -> Scatter Plot	Compare data and theoretical quantiles
6	Analysis	---	Visual inspection of the plot linearity

Note: This table is based on MS Excel workbook.

EXAMPLE OF Q–Q PLOT

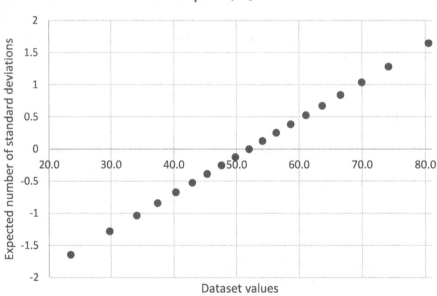

FIGURE A.3.1   Example of Q–Q plot.

Appendix 4
Nearest neighbour (NN)

Nearest neighbour interpolation is the simplest approach to interpolation. Instead of calculating an average value by some weighting criteria or generating an intermediate value based on complicated rules, this method simply determines the "nearest" neighbour value based on distance. All that the authors can say with regard to this method is to stay away entirely from this method; it is an abomination insofar as estimation is concerned. It ignores any form of rational thought and only provides some level of accuracy if the data utilised is extremely closely spaced. Consider if one has two scenarios, one in which the closest neighbour is merely 3 m away from the point to be estimated; here, one would still incur an error, however, if the grades are relatively evenly distributed in the area, this error may be of a minor nature. Now, consider that the closest neighbour is 500 m away; there is no chance on earth that this could even be in the ballpark of the value utilised.

Sample value assigned to the point of estimation:

Example

$$+4 \, g/t \quad +? \quad +10 \, g/t$$
$$A \qquad B \qquad C$$

$$A \text{ to } B = 5 \text{m}, \quad B \text{ to } C = 15 \text{m}$$

Nearest neighbour estimate for point $B = 4 \, g/t$

Advantage: quick, requires no mathematical weighting.

Disadvantage: crude, ignores spatial variability of grades and scale effect (volume-variance).

Appendix 5
Inverse distance weighting

Inverse Distance Weighting (IDW) interpolation is mathematical (deterministic), assuming closer values are more related than further values in its function.

Weighted methods (Inverse distance): Sum of weighted sample grades

Of all the distance-related estimation methods, this method is the least desirable and is set up with nothing more than thumb sucks that do not take cognizance of the orebody characteristics in a rigorous manner. Estimate = (weight$_1$ **x** sample value$_1$ + weight$_2$ **x** sample value$_2$ + ... weight$_n$ **x** sample value$_n$)

- Arithmetic mean: gives equal weighting to all samples
- Intuitively, closer samples are more correlated to the point of estimate
- Weights close samples highly
- ID Estimate = Sum of weighted sample grades
- Examples: Inverse Distance Squared (IDS) Estimate weights each sample inversely to distance^2 and divide by the sum of inverse distance^2

$$
\begin{array}{cccc}
& & +1.1 & \\
+0.5 & & B & \\
A & +? & & +1.5 \\
& C & & D \\
\end{array}
$$

$$AC = 8\,\text{m}, \quad BC = 7\,\text{m}, \quad CD = 12\,\text{m}$$

$$\text{IDW}^2 = \frac{0.5/\left(8^2\right) + 1.1/\left(7^2\right) + 1.5/\left(12^2\right)}{\left(\left(1/8^2\right) + \left(1/7^2\right) + 1/12^2\right)}$$

Advantages

- Quick
- Easy to use

Disadvantages

- Assumptions are not correct as none of the functions of distance follow a strict inverse distance rule.
- No account of clustering.
- Variogram parameters such as distances of anisotropy and scale ignored.
- Can over-predict for positively skewed/high nugget ore bodies.
- There is no indication of the error function; hence, one has no idea as to the error incurred in the estimate.

Appendix 6
Krige's relationship

Krige's relationship (otherwise known as the volume variance relationship) can be expressed as the variance of the points in the deposit (or that portion under consideration) equals the variance of the points inside each block plus the variance between the blocks in question.

Mathematical expression of Krige's relationship $\sigma^2\,\text{PIR} = \sigma^2\,\text{PIB} + \sigma^2\,\text{BB}$ where $\sigma^2\,\text{PIR}$ is the variance of the points in the resource.

$\sigma^2\,\text{PIB}$ is the variance of the points inside the averaged block.

$\sigma^2\,\text{BB}$ is the variance between the averaged blocks.

This relationship makes some assumptions that are not always explicitly declared, leading to some practitioners saying that the relationship does not appear to work under all circumstances. However, often this is due to some of the assumptions not being of a true nature. The assumptions made are as follows:

1. The distribution under consideration is Gaussian or has been transformed to Gaussian.
2. The number of points available to calculate the variance of the points in the resource is sufficient to ensure a relatively correct variance. This will impact upon the variance of the samples within the block if this is calculated utilising the semi-variogram, and hence will adversely affect the estimate of the dispersion variance of the blocks within the resource. Here, in addition, if the variance is calculated, one needs to ensure that the discretisation utilised for the calculation speaks to the number of points required in order to achieve a representative mean of the block.
3. The number of points within each averaged block is sufficient to obtain the correct average of the block. If there are insufficient samples within each block, the between block variance will be overestimated as the variance of the samples is directly a function of *n*, that is, number of samples.

Appendix 7
Semi-variogram

Semi-variograms measure spatial autocorrelation between paired samples. There may now be a question regarding how this is calculated and why it is calculated in such a way. Some of these answers can be traced by what it is that Georges Matheron was trying to achieve. This all started when Matheron chanced upon Danie G. Krige's master's dissertation and discovered Krige's famous quotation upon reading it. "It can be expected that the gold values in a whole mine will be subject to a larger relative variation than those in a portion of the mine". This meant that samples taken closer to each other are more likely to have similar values than if taken farther apart. One can only imagine the excitement that he felt at this declaration; although nowadays we think that this is merely common sense, in the 1960s and 1970s, this was far from being obvious. So, what if this relationship was modellable? One then could utilise the modelled relationship to mathematically obtain the value at an unknown position. The definition of Krige's dissertation should immediately strike one. It is not the differences here that he is talking about but rather the variability. Consider for a moment that one was considering the differences rather than the variances. It is not beyond the reach of one's imagination that if there are multiple samples to consider the relationship that, some of the differences will be positive and some will be negative, with the overall difference equalling zero. This does not satisfy the relationship that samples further apart will have a greater difference than those closer together. Hence the use of variances which under all circumstances are positive.

Appendix 8
Sampling

The objective of any sampling programme is to determine the grade, thickness, etc., of a small portion of the deposit. This can only be obtained satisfactorily by a multidisciplinary approach. Sampling data are fundamental to any mineral resource estimation exercise but are of very limited value unless correlated with all the geological parameters. The geologist must be aware of the values in his section and ensure that the geological factors are recorded to explain any anomalies. Thus, the geologist must make every effort to ensure that the sampling is based on sound geological and sampling principles. A proper understanding of the value distribution can be obtained only in this way.

EXPLAIN AND GIVE EXAMPLES OF THE RELATIONSHIP BETWEEN MINERAL VALUES AND RELATED GEOLOGICAL FEATURES/CHARACTERISTICS

Geological maps are used in planning future exploration, directing development work, and coordinating stopping. Factual geologic information is the base from which a three-dimensional image of a mineral deposit is developed. As a rule, this information is obtained from surface rock exposures, trenches, drill cores or cuttings, and underground workings. These sources provide direct observations of rocks and minerals but represent a very limited proportion of the total volume of a mineral deposit and its surroundings. Even for a well-sampled mineral deposit, the total volume of all samples could be about only one millionth of the deposit volume. Hence, a substantial interpretive component is required to develop a three-dimensional model of a mineral deposit and adjacent rocks. This interpretive component involves the interpolation of geologic features between control sites (i.e. extensions of features between known data) and may include some extrapolation (extension outward from known data).

A variety of surveys aids the interpretive process, actual survey of a position determined from a fixed survey point, geophysical and geochemical surveys that can help localise specific geologic features such as faults or particular rock types, improve confidence in the continuity of 'ore', and, in some cases, provide improved grade estimates relative to traditional assaying of geologic information is normally recorded on maps and cross sections at a scale appropriate to the aims. Property geology might be mapped at a scale of 1:5,000, whereas mineral-deposit geology might be mapped to a scale of 1:1,000 or even more detailed. Having said the above, with computer methods available nowadays, scale is becoming redundant, and the digital position of any geology can be captured as real positional coordinates and then plotted if required to any scale whatsoever. Ideally, this is the preferred methodology which provides details at the same resolution as other attributes that may be captured.

THE TYPES OF INFORMATION THAT ARE RECORDED AND DISPLAYED ON MAPS INCLUDE:

Rock types:

Rock composition influences reactivity to mineralising solutions and controls response to deformation. Rock types (including mineralised ground) are one of the most fundamental pieces of geologic information; their chemical and physical attributes and age relations provide the basic framework for understanding the geologic history of an area (e.g. pre-ore and post-ore dykes).

Faulting:

Faults disrupt and complicate the lithological record. The ages of faults are important: pre-mineralisation faults might be mineralised; post-mineralisation faults might disrupt a primary deposit and form a boundary across which it is inappropriate to extend grades for block estimation purposes.

Folding:

Folding can provide ground preparation for some types of deposits (e.g. saddle veins) and can extensively disrupt a pre-existing mineralised zone to produce a complex geometry. In the case of shear folding of a tabular deposit, mineralisation in the fold limbs can be greatly weakened, whereas a large and perhaps increased thickness can be present in the crests of folds.

Fracture/vein density and orientation:

Sites where fractures have controlled mineralisation spatial density and evidence of preferred orientation provide insight into localisation of ore and preferred directional controls.

Evidence of primary absorbency/porousness:

Porousness for mineralising fluids can be controlled by structure or by lithological character (e.g. reactive carbonate beds or breccia with substantial interconnected porosity).

Successive phases of mineralisation:

Many deposits are clearly the product of more than only one phase of mineralisation. Sorting out the characteristics of each phase (i.e. understanding the paragenesis of the mineralisation) and determining the extent to which various phases are superimposed spatially (i.e. through detailed geologic mapping) have important implications for mineral inventory estimation. For most mineral deposits, much of the detailed geologic information is obtained from logging or mapping the drill core. Consequently, it is wise to record such data in a systematic fashion that is easily adapted to a computer so that the information can be output in a variety of forms (sections, maps, correlation matrices, etc.) to assist in mineral inventory estimation.

Explain and give examples of potential consequences of sampling deviations:

a. Incorrect sampling.

Sampling is the basis of mineral resource estimation and therefore, it is necessary that it be of the highest possible standard. Incorrect sampling can affect the expected profits to be made by a mine and so cause serious financial loss because the mining of uneconomic ore which is believed to be economic due to incorrect sampling, will result in less mineral obtained than expected. Also, economic ore believed to be uneconomic will not be mined, causing loss of revenue. The grade control on a mine is primarily dependent on correct sampling information and it becomes impossible to control grade efficiently if sampling is incorrectly done. The life and future of the mine are directly dependent on having the correct information obtained from sampling results. As mineral resource estimation is dependent on the measuring and chipping of samples underground and the correct assaying of such samples, it is essential that these operations be of the highest possible standard.

"Sampling that is not of a high standard is a waste of time and money".

The position of the samples taken plays a significant role in the ultimate determination of the face grade. An entire panel was sampled in 10 cm sections and a 7 cm width in an experiment. The values along the length of the panel ranged from trace to 5,111 cm.g/t. The list of grades determined by using 2 m sample averages ranged from 224 cm.g/t to 1,747 cm.g/t. The mean based on the 10 cm spacing was 570 cm.g/t. This high variation in values is due to the log-normal distribution of gold deposits. Sample sections must be located by means of accurate measurements from survey pegs.

MARKING-OFF AND MEASURING OF SAMPLE WIDTHS

Examining and segregating wide reef channels into reef bands according to apparent quality is necessary. Sketch and correlate geological features, off-reef mining, reef not fully exposed, and reef bands intersected on the face that may affect the grade of the ore being mined. A small amount of waste rock ±1 to 2 cm above and below the top and bottom bands is marked off and included with the sample to ensure that any enrichment on reef-waste contact is included with the sample. In the case of narrow bands, it may be necessary to increase this distance to have an 8 cm sample. Accurate measuring of widths is critical as these figures are used as a multiplying factor in averaging values as well as determining tonnes broken. Any inaccurate measurements will result in overvaluation of the gold content and overestimation of the tonnage broken. In the case of very narrow, highly carbonaceous reefs, more of the softer reef and less of the harder waste rock is sampled. For this reason, the sample widths should be measured after the sample has been taken.

For example, a sample of 12 cm wide has been erroneously measured as 13 cm.

The error in the over-measurement of the width is $((1 \div 12) \times 100) = 8.3\%$.

If the assay value returned for the sample is 50 g/t:

True value $= 12 \times 50 = 600$ cm.g/t

False value $= 13 \times 50 = 650$ cm.g/t

The error in the overvaluation of the sample is $((50 \div 600) \times 100) = 8.3\%$.

The effect in the mining grade if the stope width is 100 cm:

True grade $= 600 \div 100 = 6.0$ g/t

False grade $= 650 \div 100 = 6.5$ g/t

The above example clearly shows how an error of 1 cm in the measurement of a sample can introduce an error of 8% in the valuation of a narrow reef.

b. Sampling the incorrect reef.

Sampling the incorrect reef can affect the expected profits to be made by a mine and so cause serious financial loss because the mining of uneconomic ore which is believed to be economic due to incorrect sampling will result in less mineral obtained than expected. Also, economic ore believed to be uneconomic will not be mined, causing loss of revenue.

Sampling the incorrect reef will also have a negative influence on the integrity of the database if the reef was not captured correctly.

c. Assuming the right reef and not sampling it.

Sample every exposure of any mineral deposit and record the results on plans, to determine which areas in the mine can be mined at a profit and to determine the life of the mine.

d. Lack of verification and checking of sample data.

Lack of verification or checking of data can and will result in errors within estimation. Some of the factors and resultant errors are listed in Table A.8.1.

e. Over-/under-evaluation.

Several factors contribute to over- and/or under-evaluation and are summarised in Table A.8.2.

As seen above, over- and under-evaluation can coincide on both sides.

To eliminate over- and under-evaluation, the following rules must be adhered to:

1. Samplers with a high degree of integrity are to be used.
2. A reliable and tested mineral resource estimation protocol must be in place.
3. Ensure that regular planned and unplanned task observations are performed on samplers and geologists core logging.
4. Double-check information captured in the database.
5. Do regular checks by checking raw data inputs into electronic sampling systems.
6. Print value plots and check for sampling points that are suspiciously placed or plotting into the solid.
7. Do regular audits on Assay laboratories.
8. Regularly do planned task observations on the sampling store to ensure that samples are conveyed, stored, and transported to Assay laboratories without being compromised.

TABLE A.8.1
Factors and result errors

Factor	Result
Lack of checking laboratory QA/QC results.	Potential biases go unnoticed. Incorrect or erroneous values captured.
Lack of verification of core logging.	Stratigraphy incorrectly logged; sample widths incorrectly captured. Reef captured as waste or vice versa. Sample ticket numbers incorrectly assigned.
Lack of verification of channel sampling.	Data captured on incorrect reef. Reef captured as waste or vice versa. Sample ticket numbers incorrectly assigned. Sample widths incorrectly captured.
Lack of checking assay results.	Ticket numbers incorrectly assigned. Suspicious values never queried. Waste assigned reef values and vice versa. Suspicious bullion corrections not noted. Multi-mineral assays with missing values not noticed, with a potential trace value being automatically being assigned.

TABLE A.8.2
Reasons for over and/or under-evaluation

Over-evaluation	Under-evaluation
Incorrect sampling methods	Incorrect sampling methods
Incorrect sampling measuring	Incorrect sampling measuring
Incorrect estimation methods	Incorrect estimation methods
Database that is not trustworthy	Database that is not trustworthy
Incorrect values from assay labs	Incorrect values from assay labs
Biasness	Biasness
Mineral salting	Volume salting
Incorrect geological interpretations	Incorrect geological interpretations
Incorrect measuring	Incorrect measuring
Poor sampling coverage	Poor sampling coverage

9. All geological disturbances must be reported to Geology for investigation without delay.
10. A close relationship between geology and sampling must exist to ensure effective and correct interpretation of all geological disturbances.
11. Report any suspicious over or under-monthly measuring.

f. Influence on mining decisions.

Incorrect sampling can affect the expected profits to be made by a mine and cause a severe financial loss because the mining of uneconomic ore, which is believed to be economic due to incorrect sampling, will result in

less mineral obtained than expected. Also, economic ore believed to be uneconomic will not be mined, causing a loss of revenue.
  g. False ore reserve declarations.
      Resource/reserve declarations should be based on sound geological and economic certainty levels.

GOVERNING PRINCIPLES

Guidelines were written considering best industry practices and the Committee for Mineral Reserves International Reporting Standards (CRIRSCO) mission to protect investors and maintain the integrity of the securities markets. All investors, whether large institutions or private individuals, should have access to specific basic facts about an investment prior to purchasing or selling it.

For example, the South African Code for Reporting of Exploration Results, Mineral Resources and Mineral Reserves (SAMREC) code requires South African mining companies to disclose meaningful financial and other information to the public, which provides a shared pool of knowledge for all investors to use to judge for themselves if a company's securities are a worthwhile investment. The public can make sound investment decisions through the steady flow of timely, comprehensive, and accurate information. To meet the SAMREC code requirements for disclosure, a mining company must make available all information, whether it is positive or negative, that might be relevant to an investor's decision to buy, sell, or hold the security.

TRANSPARENCY, MATERIALITY AND COMPETENCE

The main principles governing the development and application of the SAMREC code are transparency, materiality, and competence.

- Transparency requires that the reader of a public report be provided with sufficient information, the presentation of which is unambiguous, to understand the report and not be misled.
- Materiality requires that a public report contains all the relevant information that investors and their professional advisers would require and reasonably expect to find in a public report to make a reasoned and balanced judgement regarding the exploration results, mineral resources or mineral reserves being reported.
- Competence requires that the public report be based on work that is the responsibility of suitably qualified and experienced persons subject to an enforceable professional code of ethics and rules of conduct.

The following additional principles must also be considered:

- Consistency between financial and technical reports:
    Financial reports consider mineral resources and ore reserves based on assumptions concerning commodity prices, exchange rates, and other

parameters of significance. To be detailed, technical and financial information should be published on a comparable basis.

- Consistency between financial markets:
    Global companies can only achieve transparency if the information is reported consistently in all financial markets. Only then can the information supplied to all investors be identical, clear, and unambiguous.

A competent person must sign off ore reserves. Complaints made in respect of the professional work of a competent person will be dealt with under the disciplinary procedures of the professional organisation to which the competent person belongs.

h. Life of mine.

The life of a mine does not start the day that production begins but many years before, when the company sets out to explore for a mineral deposit. A lot of time and money is spent simply looking for, locating, and quantifying a promising mineral occurrence. Not many will be found, and not many of the ones found will have the potential to become mine. Spending five to ten years searching for a mineable deposit is not unusual.

DEFINITION

The time in which, through the employment of the available capital, the ore reserves – or such a reasonable extension of the ore reserves as conservative geological analysis may justify – will be extracted.

After preliminary exploration has been completed, the next step in the life cycle of a mine is resource definition. The processes in reserve definition are similar to preliminary exploration, and there is considerable overlap between the two phases. However, in resource definition, the exploration company have reason to believe that their property contains a mineral reserve. Therefore, the company will spend more money on intensive and technical exploration techniques.

The resource definition phase not only involves further analysis of the size and grade of the mineral reserve but also uses engineering and geotechnical studies to evaluate the mining method and estimate how much it will cost to extract the ore, given the geology of the deposit. A feasibility study is published at the end of this process, and the ore deposit may be deemed uneconomic or economic.

i. Mine call factor (MCF).

The Institute of the Mine Surveyors of South Africa defines it as follows:
The MCF is a ratio expressed as a percentage, which the specific mineral product accounted for in "recovery plus residue" bears to the related product "called for" by the mine's measuring and valuation methods.

TABLE A.8.3
Apparent metal loss versus real metal content loss

Apparent Gold Loss	Real Gold Loss
Inaccurate sampling	Inefficient mining methods and layouts
Incorrect densities used	Poor quality of sweepings
Over/under measurements of Stoping Width and area mined	Reef left in hang wall and footwall Unaccounted
Inaccurate valuation methods applied	Old tonnage and mud not cleaned up
Incorrect silver correction	Theft
Inaccurate assaying of samples	Losses during tramming and hoisting
	Reef to waste
	Losses due to leaking pipes in plant
	Losses due to belt spillages that are not cleaned up
	Plant losses due to incorrect dosing

Suppose the sampling, assaying and tonnage measurements in a mine and plant are perfect, and none of the mineral content is lost at any stage during the handling and processing. In that case, the MCF should theoretically be 100%.

An MCF of 90% is generally considered acceptable, but unfortunately, this measure is below the acceptable norm in many mines.

Numerous factors contribute to keeping the MCF below 100%, either *apparent losses* or *actual losses*. The real gold loss should be traceable; if lost during mining operations, it should be found underground. It can only be ascribed as apparent gold loss if it cannot be found underground.

Some of these actual losses are small and become constant over time. For example, during the clean-up process, gold is lodged into the cracks and crevices and spilt along tramming and hoisting routes. This gold is not efficiently recovered during normal sweeping operations but only when final stripping and vamping are done. Similarly, the gold plant absorbs gold throughout its life, and this small amount is only recovered when the plant is dismantled.

The MCF can also be split into an underground and surface measurement by determining the shaft call factor (SCF) and the plant call factor (PCF).

PCF = Gold accounted for by the plant/Gold estimated over the belt

SCF = Gold estimated over the belt/gold called for

Therefore MCF = PCF × SCF

Areas of impact on the MCF are, therefore, the metallurgical plant, assay laboratory, sampling methodology, survey measuring accuracy, mining methodology, and the shafts ore accounting methodology. Fluctuations in the monthly mined value and the time lag of ±14 days between production and milling periods also result in variations in the MCF. Therefore, a six-month moving average should be considered a more realistic MCF. If the MCF is consistently below 80%, only should investigate why, where, and how the gold loss occurs. Conversely, an unusually high average MCF indicates that the mine's measuring, sampling, and valuation methods must be investigated.

The MRM department's responsibility is to ensure that quality tonnes are sent from the stope face to the plant.

Key focus areas:

- Quality sweeping
- Managing fragmentation
- Minimising water usage
- Good housekeeping in the haulages, cross cuts, and stopes
- Effective blasting barricades
- Booking, reporting, Short Interval Controls
- Correct tramming: reef to reef and waste to waste
- Manage mining mix
- Increase quality volume
- Stoping Width (SW) controls
- Elimination of off-reef mining
- Reef fully exposed

THE SIGNIFICANCE OF THE VARIABLE COMPONENTS OF THE MINE CALL FACTOR IN MAINTAINING THE GRADE

The reason for a low MCF is usually due to errors in estimating the quantity and grade of ore which are calculated from survey measurements and underground sampling.

Allowance should therefore be made for known losses as well as unrecorded losses of the mineral, that is, those which are unavoidable and those which are due to negligence. Normally, grade problems are accompanied by a low MCF which is due to losses of the mineral somewhere along the line, and a systematic examination of the following variable components is necessary to trace the cause:

SURVEYOR'S MEASUREMENTS AND CALCULATIONS

a. Stope measurements

Because all measurements are regulated by law (Chapter 17 of the Mine Health and Safety Act, Act 29 of 1996 in South Africa) and the contractor's payments are calculated from these measurements, it is unlikely that any serious errors will occur because of inaccurate measurements.

b. Stope widths

Although errors in the measurement of stope widths must affect the calculations of tonnage broken and the grade of tonnage in terms of g/t, these errors do not affect the estimation of the total quantity of gold if the additional rock broken is waste. Consequently, errors in widths have little or no effect on the MCF if the gold called for is based on survey tonnage broken. However, having said that, all tonnage to the slime dams contain mineral content of interest thus it can be said that tonnage from an excessive stoping width goes into the plant at zero grade and comes out of the plant with grade and thus steals some of the grade that could have been recovered.

c. Development

Strict control of the tramming of development rock as ore or waste must be exercised to ensure that all ore is trammed as such. Should development ore be reported as ore being "trammed as ore" when, in fact, it has been treated as waste, the mineral content therefrom will be erroneously included in the amount of content called for, and therefore the MCF will be lower? On the other hand, when waste tonnage is trammed as ore, the MCF will be unaffected because the content called therefrom will be zero (zero g/t × tonnes = 0 g). This will, however, result in the increase in shortfall tonnages because waste tonnages are not included in flow sheet calculations.

d. Ore from dumps

Where ore is being taken from surface dumps, it is necessary to obtain an accurate estimate of the quantity involved. It is recommended that densities of broken rock and/or slimes are checked frequently. Sampling methods should also be investigated to ensure that sampling values are representative of the values inside the dump.

CURRENT SAMPLING OF ORE SENT TO MILL

Failure to maintain high sampling and assaying standards can have a profound effect on the MCF. Efficient supervision is necessary underground and on surface to ensure that samplers comply with standard sampling measurement and calculation procedures.

Rock packed underground as waste

Waste sorted and packed underground is measured and plotted on stope plans. Waste packs should be sampled regularly because the traditional zero value allocated to them proved to be incorrect because of human errors during packing and reef fines which are blasted into the packs.

Rock sorted as waste on surface

The amount sorted as waste on surface can easily be over-estimated when cars are not properly filled. Sampling of sorted waste is liable to personal bias, and such work should be undertaken by trained personnel.

Ore picked on surface from waste

Care must be taken in the estimation of the value and quantity of ore milled from reef-picking plants.

TIPPING OF ORE AND ALLOCATION OF CAR FACTORS

Every precaution should be taken to ensure that ore is not inadvertently or deliberately tipped as waste during its passage to the mill. This problem is normally due to a lack of supervision.

Numbered steel washers scattered in waste development can serve as a check to identify problem areas. When waste is trammed improperly as ore, these washers are recovered from magnets fixed over the conveyor belts at the mill. Their identification numbers will inform management who are responsible for the dilution of ore. Car factors and the filling of cars must be checked periodically, where car tallies are used in calculating the mineral content called for.

LOSSES OR THEFT IN PLANT

Once the initial absorption of mineral has taken place in a plant, very little of the mineral is lost thereafter. Strict security measures will ensure that losses due to theft are extremely rare occurrences.

ASSAY BIAS AND ALLOWANCE FOR SILVER CONTENT IN GOLD ASSAYING

The mine assay procedures must be investigated because it can easily cause a consistently poor MCF. Allowance for silver must be checked frequently, and the relationship between gold grades and silver modelled as the ratio does not adhere to the 10% thumb suck applied at many mines. The same can be said of the relationship between minerals on a multi-mineral deposit.

Appendix 9
Core logging

The purpose of core logging is to:

- Provide geological information such as position and grade of the ore and position of faults.
- Provide warning of the presence of water and/or gas ahead of development.
- Provide drain holes for waste mine water and pathways for cables.
- Provide information on structure, sedimentology, geochemistry, stratigraphy, etc.
- Evaluate the orebody (reef) ahead of mining.
- Increase ore reserve (long exploration holes).

The role of a geologist in drilling and core logging is to:

- Determine the need – what to drill through and why (soil, rock, dump, and mine workings),
- Clarify the purpose,
- What will be logged and/or sampled,
- Drill style and rig,
- Sample type needed,
- Create and/or scrutinise contract,
- Monitor and supervise drill operation – ensure compliance – technical, legal, and environmental,
- Collect information to substantiate drill reports subsequent,
- Interrogate, learn from and support drilling crew,
- Draw sections and plans,
- Change the plan if necessary – for example, drill styles, sample types, holes, terminate, or proceed decisions,

MECHANISM FOR DRILLING CAN BE CLASSIFIED INTO THREE CATEGORIES

1. Crushing
2. Shearing
3. Abrading

All produce either mud, silt, sand, chips, or core. The most common method is Diamond drilling and Reverse Circulation. Cutting accomplished by small diamonds

FIGURE A.9.1   Example of drill bits.

TABLE A.9.1
Common drill diameters

ID	Outside Diameter (mm)	Internal Diameter (Core) (mm)
AQ	47.75	27.1
BQ	59.69	36.52
NQ	75.44	47.75
AXT	47.75	32.66

embedded into a steel drill bit. Most popular for geological, mining, and prospecting work but most expensive. Various sizes of holes and cores.

Some of the most common drill core diameters are shown in Table A.9.1

CALCULATE MINIMUM WEIGHT FOR SAMPLE REQUIRED FROM DRILL CORE

Density $(g/cm^3) = m\ (g)/v\ (g/cm^3)$

Cylinder volume $= \pi \cdot r^2 \cdot h$ where h is the length of full core required

Say we need 250 g as a minimum weight (confirm with your assay lab)

- AXT size radius = 1.633 cm
- Relative Density of bulk rock = 2.75 g/cm^3
- The volume (v) of core required is = 250 g/2.65 g/cm^3 = 90.9 cm^3
- Therefore, the length of AXT size core required is h = 90.9 cm^3/(3.1459 × 1.633 cm^2) = 10.85 cm
- Obviously, the figure needs to double if you are using half cores for assaying

Appendix 10
Calculating the economic breakeven

As intimated previously, the economic breakeven grade is the grade at which the mine neither operates at a profit nor a loss. In order to calculate this parameter requires some assumptions to be made. Due care must be taken that these are of a realistic nature as subsequent calculations such as the cut-offs, proportions available to mine and change of support have this calculation as an underlying basis. The assumptions are as follows:

- The tonnage that will be mined on the period under review.
- The mineral/minerals prices that will occur during such period.

TABLE A.10.1
Calculations of process flow

					Process Flow	
	%	Tonnes	Grade	Content	Remarks	Direction of calcs
			(g/t)	(kg)		
Recovery	97.5	150,000	1.282	192.4	Total working cost by tone = Gold required	
Residue	2.5	150,000	0.033	4.9		
Total to mill		150,000	1.3	197.3		
Mine call factor	87.0	150,000	1.5	226.8		
Ore hoisted		150,000	1.5	226.8		
Dev to mill	2.5	3,759	1.5	5.6		
Waste dev to mill	3.0	4,500		0.0		
Stope ore to surface		141,741	1.6	221.1		
Tonnage discrepancy	5.1	7,575				
Total measured from stopes		134,166	1.6	221.1		
Other sources stoping	5.6	8,419	0.5	4.2		
Measured stoping ore		125,747	1.7	216.9	Breakeven grade highlighted	

- The percentage of other sources stoping.
- The percentage of reef from development and an estimate of the associated grade.
- The mine call factor.
- The percentage of residues from plant.
- The tonnage discrepancy.
- Total working cost for tonnage mined for period under review.

Below is an example of a breakeven grade calc. The red is the final breakeven grade; the blues are the assumptions alluded to previously. All the rest are calculations based on those assumptions, starting with gold recovered (which is what we get our revenue from) down to what is required grade from stoping. As you may notice, the calculations are an ore flow in reverse order.

Appendix 11
Hands-on practical python codes

Index

Note: **Bold** page numbers refer to tables; *italic* page numbers refer to figures and page numbers followed by "n" denote endnotes.

Printed in the United States
by Baker & Taylor Publisher Services